AF576063

Composite Materials in Design Processes

Composite Materials in Design Processes

Editor

Giangiacomo Minak

MDPI • Basel • Beijing • Wuhan • Barcelona • Belgrade • Manchester • Tokyo • Cluj • Tianjin

Editor
Giangiacomo Minak
Università di Bologna
Italy

Editorial Office
MDPI
St. Alban-Anlage 66
4052 Basel, Switzerland

This is a reprint of articles from the Special Issue published online in the open access journal *Applied Sciences* (ISSN 2076-3417) (available at: https://www.mdpi.com/journal/applsci/special_issues/composite_materials_design).

For citation purposes, cite each article independently as indicated on the article page online and as indicated below:

LastName, A.A.; LastName, B.B.; LastName, C.C. Article Title. *Journal Name* **Year**, *Volume Number*, Page Range.

ISBN 978-3-0365-0152-9 (Hbk)
ISBN 978-3-0365-0153-6 (PDF)

Contents

About the Editor

Giangiacomo Minak Associate Professor of Mechanical Design and Machine Elements. Ph.D. in Nuclear Engineering at the University of Bologna, 1999. His main research field is the mechanical behavior of composite materials, particularly under impact and fatigue loading for the design of lightweight vehicles (e.g., Emilia 4, winner of the America Solar Challenge 2018). A second interest is the "impossible design", i.e., additive manufacturing design.

Editorial

Special Issue "Composite Materials in Design Processes"

Giangiacomo Minak

Department of Industrial Engineering (DIN), Alma Mater Studiorum, Università di Bologna, 47121 Forlì, Italy; giangiacomo.minak@unibo.it

Received: 26 November 2020; Accepted: 1 December 2020; Published: 3 December 2020

Composite materials have been used in design since antiquity, as the description of the Ulises' arch in the Odyssey suggests [1]. The great advantage provided by the use of composite materials in the design process is that it allows tailoring the mechanical properties of the components, in order to obtain the highest specific strength or stiffness and, consequently, reduce the overall weight. The possible combinations of matrix, reinforcement, and technologies, on the one hand, provide many more options to the designer and, on the other hand, widen the fields that need to be investigated to obtain all the information requested for a safe design.

This Special Issue contains a variety of approaches aimed to draw directions for the designers of applications characterized by different technology readiness levels, at different dimensional scales and technological process phases.

Design of the material: A number of papers may be categorized in this way, from different points of view. In [2], the toughening of the matrix through polymeric nano veils is described. This is a popular research topic because employing low-cost electrospinning technology is easily possible to obtain non-woven nanostructured veils starting from different liquid polymers. The research activity is based on the relatively recent review presented in [3], and since then, different research groups have provided results in this field [4–7].

In [8], the concept is widened, since the epoxy resin is reinforced by a microcapsule system to achieve a self-healing goal. In this case, there are no reinforcing fibers, because the role of the epoxy resin is functional (i.e., electrical insulation) and not structural; nevertheless, the authors obtained a very promising repair efficiency and rate.

The papers [9,10] consider two different aspects of interleaving viscoelastic materials between long fiber reinforced plastics layers. In fact, while [9] focuses the attention on the positive effect of macroscopic viscoelastic elements on the slamming damage in Glass Fiber Reinforced Plastics (GFRP), in [10], thin viscoelastic layers are interleaved with Carbon Fiber Reinforced Plastics (CFRP) plies in order to improve the damping properties of the laminate.

Differently from the previously mentioned ones that focus on matrix properties modification, the works [11,12] are more related to fibers. In [11], several methods for measuring the fiber content in naval applications are shown. This is of paramount importance for design purposes, since the fiber fraction is the quantity most of the mechanical properties are most sensitive to, according to the rule of mixtures. On the other hand, in [12], an analytical model is proposed and experimentally validated to describe the shear behavior of fabrics with different weave patterns, in which tension-shear coupling is considered. This is done under large shear deformation.

Improvement of the technological processes: Dealing with composite materials, it is not possible to separate the material properties from the technological production processes. In [13,14], two dry fiber techniques are studied. In particular, [13] shows the numerical simulation of the channel distribution in a liquid resin infusion using an approach similar to the one reported in [15]. Differently, [14] presents an experimental activity on the resin transfer molding process following the previous activities on the topic [16,17] of the research group. The papers [18,19] focus on the electric deposition of composite coatings and coupled waterjet and laser surface treatments.

Structural design methods: Even if just one paper belongs to this category [20], it is described separately because it is the one dealing with applications at TRL6 or higher and contains numerical and experimental analysis. Moreover, a synthesis of a complete design and fabrication of an automotive part, in particular the photovoltaic panel of a solar vehicle, is presented. The general requirements of this class of vehicles are described in [21], while the design process for the structure and some mechanical components can be found in [22–25]. The paper contained in this Special Issue deals with a multi-objective design optimization in which not only the lamination sequence but also the topology of the component is modified to obtain optimum performances.

Research directions: This Special Issue touched several hot research topics showing that, to improve how designers can use composite materials, different activities are needed. New functional properties may be sought for maintaining or improving material mechanical performances; while low cost, high-volume processes would open a wider market for composite components, so this is a further wide research field; finally, general topological optimization methods for layered materials will allow us to reduce the weight in the most advanced sport of aerospace applications further.

Funding: This research received no external funding.

Acknowledgments: The Guest Editor would like to thank all authors, the many dedicated referees, the editor team of Applied Sciences, and especially Snežana Repić (Assistant Managing Editor) for their valuable contributions, making this Special Issue a success.

Conflicts of Interest: The author declares no conflict of interest.

References

1. Verity, A.; Allan, W. *Homer. The Odyssey*; Oxford University Press: Oxford, UK, 2016; Chapter XXI.
2. Saghafi, H.; Palazzetti, R.; Heidary, H.; Brugo, T.M.; Zucchelli, A.; Minak, G. Toughening Behavior of Carbon/Epoxy Laminates Interleaved by PSF/PVDF Composite Nanofibers. *Appl. Sci.* **2020**, *10*, 5618. [CrossRef]
3. Saghafi, H.; Fotouhi, M.; Minak, G. Improvement of the Impact Properties of Composite Laminates by Means of Nano-Modification of the Matrix—A Review. *Appl. Sci.* **2018**, *8*, 2406. [CrossRef]
4. Saghafi, H.; Moallemzadeh, A.R.; Zucchelli, A.; Brugo, T.M.; Minak, G. Shear mode of fracture in composite laminates toughened by polyvinylidene fluoride nanofibers. *Compos Struct.* **2019**, *227*, 111327. [CrossRef]
5. Saghafi, H.; Minak, G.; Zucchelli, A.; Brugo, T.M.; Heidary, H. Comparing various toughening mechanisms occurred in nanomodified laminates under impact loading. *Compos. B Eng.* **2019**, *174*, 106964. [CrossRef]
6. Barzoki, P.K.; Rezadoust, A.M.; Latifi, M.; Saghafi, H.; Minak, G. Effect of nanofiber diameter and arrangement on fracture toughness of out of autoclave glass/phenolic composites—Experimental and numerical study. *Thin Walled Struct.* **2019**, *143*, 106251. [CrossRef]
7. Fotouhi, M.; Fragassa, C.; Fotouhi, S.; Saghafi, H.; Minak, G. Damage characterization of nano-interleaved CFRP under static and fatigue loading. *Fibers* **2019**, *7*, 13. [CrossRef]
8. Wang, Y.; Li, Y.; Zhang, Z.; Zhao, H.; Zhang, Y. Repair Performance of Self-Healing Microcapsule/Epoxy Resin Insulating Composite to Physical Damage. *Appl. Sci.* **2019**, *9*, 4098. [CrossRef]
9. Townsend, P.; Suárez Bermejo, J.C.; Pinilla, P.; Muñoz, N. Is the Viscoelastic Sheet for Slamming Impact Ready to Be Used on Glass Fiber Reinforced Plastic Planning Hull? *Appl. Sci.* **2020**, *10*, 6557. [CrossRef]
10. Troncossi, M.; Taddia, S.; Rivola, A.; Martini, A. Experimental Characterization of a High-Damping Viscoelastic Material Enclosed in Carbon Fiber Reinforced Polymer Components. *Appl. Sci.* **2020**, *10*, 6193. [CrossRef]
11. Han, Z.; Jeong, S.; Noh, J.; Oh, D. Comparative Study of Glass Fiber Content Measurement Methods for Inspecting Fabrication Quality of Composite Ship Structures. *Appl. Sci.* **2020**, *10*, 5130. [CrossRef]
12. Wang, Y.; Zhang, W.; Ren, H.; Huang, Z.; Geng, F.; Li, Y.; Zhu, Z. An Analytical Model for the Tension-Shear Coupling of Woven Fabrics with Different Weave Patterns under Large Shear Deformation. *Appl. Sci.* **2020**, *10*, 1551. [CrossRef]

13. Magalhães, G.M.C.; Fragassa, C.; Lemos, R.L.; Isoldi, L.A.; Amico, S.C.; Rocha, L.A.O.; Souza, J.A.; dos Santos, E.D. Numerical Analysis of the Influence of Empty Channels Design on Performance of Resin Flow in a Porous Plate. *Appl. Sci.* **2020**, *10*, 4054. [CrossRef]
14. Falaschetti, M.P.; Rondina, F.; Zavatta, N.; Gragnani, L.; Gironi, M.; Troiani, E.; Donati, L. Material Characterization for Reliable Resin Transfer Molding Process Simulation. *Appl. Sci.* **2020**, *10*, 1814. [CrossRef]
15. Poodts, E.; Minak, G.; Dolcini, E.; Donati, L. FE analysis and production experience of a sandwich structure component manufactured by means of vacuum assisted resin infusion process. *Compos. B Eng.* **2013**, *53*, 179–186. [CrossRef]
16. Rondina, F.; Taddia, S.; Mazzocchetti, L.; Donati, L.; Minak, G.; Rosenberg, P.; Bedeschi, A.; Dolcini, E. Development of full carbon wheels for sport cars with high-volume technology. *Compos. Struct.* **2018**, *192*, 368–378. [CrossRef]
17. Poodts, E.; Minak, G.; Mazzocchetti, L.; Giorgini, L. Fabrication, process simulation and testing of a thick CFRP component using the RTM process. *Compos. B Eng.* **2014**, *56*, 673–680. [CrossRef]
18. Lai, L.; Li, H.; Sun, Y.; Ding, G.; Wang, H.; Yang, Z. Investigation of Electrodeposition External Conditions on Morphology and Texture of Ni/SiC*w* Composite Coatings. *Appl. Sci.* **2019**, *9*, 3824. [CrossRef]
19. Chen, X.; Li, X.; Wu, C.; Ma, Y.; Zhang, Y.; Huang, L.; Liu, W. Optimization of Processing Parameters for Water-Jet-Assisted Laser Etching of Polycrystalline Silicon. *Appl. Sci.* **2019**, *9*, 1882. [CrossRef]
20. Pavlovic, A.; Sintoni, D.; Fragassa, C.; Minak, G. Multi-Objective Design Optimization of the Reinforced Composite Roof in a Solar Vehicle. *Appl. Sci.* **2020**, *10*, 2665. [CrossRef]
21. Minak, G.; Fragassa, C.; de Camargo, F.V. A Brief Review on Determinant Aspects in Energy Efficient Solar Car Design and Manufacturing. In *Sustainable Design and Manufacturing (SDM) 2017*; Campana, G., Ed.; Smart Innovation, Systems and Technologies; Springer: Cham, Switzerland, 2017; Volume 68, pp. 847–856.
22. Minak, G.; Brugo, T.M.; Fragassa, C.; Pavlovic, A.; Zavatta, N.; De Camargo, F. Structural Design and Manufacturing of a Cruiser Class Solar Vehicle. *J. Vis. Exp.* **2019**, *143*, e58525. [CrossRef]
23. Fragassa, C.; Pavlović, A.; Minak, G. On the structural behaviour of a CFRP safety cage in a solar powered electric vehicle. *Compos. Struc.* **2020**, *252*, 112698. [CrossRef]
24. Pavlović, A.; Sintoni, D.; Minak, G.; Fragassa, C. On the modal behaviour of ultralight composite sandwich automotive panels. *Compos. Struc.* **2020**, *248*, 112523.
25. Minak, G.; Brugo, T.M.; Fragassa, C. Ultra-High-Molecular-Weight Polyethylene Rods as an Effective Design Solution for the Suspensions of a Cruiser-Class Solar Vehicle. *Int. J. Pol. Sci.* **2019**, 8317093. [CrossRef]

Publisher's Note: MDPI stays neutral with regard to jurisdictional claims in published maps and institutional affiliations.

Article

Toughening Behavior of Carbon/Epoxy Laminates Interleaved by PSF/PVDF Composite Nanofibers

Hamed Saghafi [1,2,*], Roberto Palazzetti [3,*], Hossein Heidary [1], Tommaso Maria Brugo [4], Andrea Zucchelli [4] and Giangiacomo Minak [4]

1 Department of Mechanical Engineering, Tafresh University, Tehran Road, Tafresh 3951879611, Iran; hosseinheidary@gmail.com
2 New Technologies Research Center (NTRC), Amirkabir University of Technology, Tehran 1591633311, Iran
3 DMEM Department, University of Strathclyde, Glasgow G1 1XJ, UK
4 Department of Industrial Engineering (DIN), Alma Mater Studiorum, Università di Bologna, 40136 Bologna, Italy; tommasomaria.brugo@unibo.it (T.M.B.); a.zucchelli@unibo.it (A.Z.); giangiacomo.minak@unibo.it (G.M.)
* Correspondence: saghafi@tafreshu.ac.ir (H.S.); roberto.palazzetti@strath.ac.uk (R.P.)

Received: 30 July 2020; Accepted: 12 August 2020; Published: 13 August 2020

Featured Application: The present findings may find application in manufactured composite material for engineering purposes, load-bearing parts, and structural components.

Abstract: This paper presents an investigation on fracture behavior of carbon/epoxy composite laminates interleaved with electrospun nanofibers. Three different mats were manufactured and interleaved, using only polyvinylidene fluoride (PVDF), only polysulfone (PSF), and their combination. Mode-I and Mode-II fracture mechanics tests were conducted on virgin and nanomodified samples, and the results showed that PVDF and PSF nanofibers enhance the Mode-I critical energy release rate (G_{IC}) by 66% and 51%, respectively, while using a combination of the two registered a 78% increment. The same phenomenon occurred under Mode-II loading. SEM micrographs were taken, to investigate the toughening mechanisms provided by the nanofibers.

Keywords: composite laminates; nanofibers; fracture; polyvinylidene fluoride; polysulfone

1. Introduction

Carbon-fiber-reinforced polymer composites (CFRP) are applied widely in various industries, such as electronics, construction, and aeronautics. Among different resins, epoxy is the most frequently used because of its good mechanical properties, suitable fatigue resistance, and low shrinkage while curing. On the other side, its highly crosslinked structure leads to brittleness and thus to poor resistance to crack propagation [1,2]. Among the several methods that have been presented during the years to increase the fracture toughness of carbon/epoxy laminates [3–6], interleaving polymers [7–9], in the form of particles, films, or nanofibrous mats [10–16], has proved to be one of the most effective. In particular, nanofibrous mats have been found to be a suitable choice because of their high porosity (which lead to rapid penetration of epoxy) and the strengthening effects they are able to provide.

Literature reviews on nanofibers reinforcing composites are wide, and many polymers, such as polyvinylidene fluoride (PVDF) [17–20], polyvinyl butyral (PVB) [21–23], polysulfone (PSF) [10,11,24], Nylon [25–32], phenoxy [33,34], and carbon [35–37] nanofibers, have been used to enhance composites' mechanical properties. Saghafi et al. [20] Showed that PVDF nanofibers can increase Mode-I fracture toughness by about 43%, while another study [19] in this field had completely reverse outcomes. The considerations showed that the main reason was a non-suitable curing process and the high thickness of the nano-mat in the second study. As seen, some limited study was also conducted regarding the effect of PSF nanofibers on fracture behavior of nanomodified laminates. For instance, Li et al. [10] used PSF nanofibers and PSF/carbon nanotube (CNT) hybrid nanofibers for increasing Mode-II energy release rate (G_{IIC}) of carbon/epoxy laminate. According to results, PSF and the best combination of hybrid nanofibers (PSF + 10%wt of CNT) improved G_{IIC} by about 11% and 50%, respectively.

The interesting matter in this regard is the toughening mechanisms that lead to improved properties when polymeric nanofibers are interleaved. (1) Fiber bridging: When nanofibers do not melt during curing cycle, they bridge the two layers they are interleaved between, thus hindering fracture propagation [38]. (2) Phase separation: Some nanofibers, such as polycaprolactone (PCL), due to the heat provided during the curing process, change shape to spherical particles and distribute in the matrix during curing, increasing fracture toughness due to crack deflections [11]. (3) Some other thermoplastic polymers, such as PVDF, melt and mixed with epoxy during curing, due to high porosity of the mat, and a plastic zone is produced in front of crack tip, capable of absorbing energy during loading [17].

Interleaving nanofibers that can act different toughening mechanisms is an interesting topic, and this is what this paper means to present. Recently, Zheng et al. [39] used a combination of nylon nanofibers and PCL film as interleave to increase the interlaminar fracture energy of carbon/epoxy laminates. The results demonstrated a synergistic effect; for instance, Mode-I fracture tests proved that fracture toughness for the laminates interleaved by nylon and PCL, separately, were enhanced by 30% and 50%, respectively, while a remarkable increase of 110% occurred for the laminates interleaved by nylon/PCL. In the present study, the effect of mixing two other mechanisms, i.e., phase separation and plastic zone, is considered. For this aim, electrospun PSF, PVDF, and PSF/PVDF nanofibers were produced separately and interleaved between carbon/epoxy laminate. Then, Mode-I and Mode-II fracture tests were conducted to investigate their effect. For deeper investigation, SEM pictures were also taken to find out toughening mechanism.

2. Materials and Methods

Electrospinning is a technique that uses a high-potential electrostatic field to produce fibers in scale of nano and micro. The machine used to produce the nanofibers is made of (1) a high-voltage source with positive or negative polarity, (2) a syringe pump with Teflon tubes to carry the solution to needles, and (3) a conductive collector, in the form of a rotating drum. The electrospinning process is schematically shown in Figure 1. In the following subsections, further information regarding the applied materials and electrospinning parameters, such as voltage and injection rate, are presented.

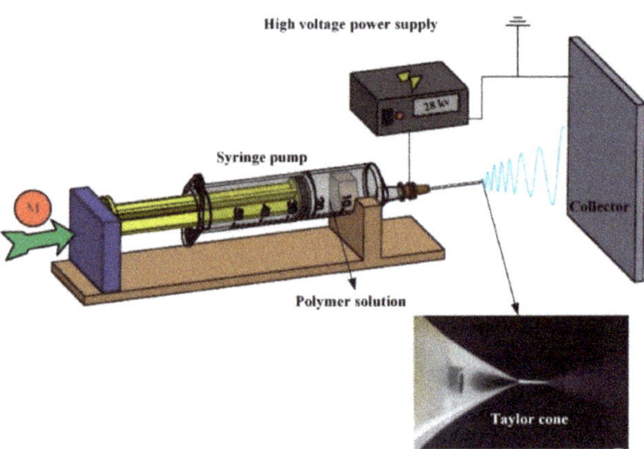

Figure 1. Schematic picture of producing nanofibers by using the electrospinning process.

2.1. Polymers

Polysulfone (Udel® 3500) and polyvinylidene fluoride (Solef® 6008) polymers in the form of pellets and powder, respectively, were supplied by Solvay Specialty Polymers. Their properties are presented in Table 1. Acetone and N, N-Dimethylacetamide (DMAc) and Dimethyl sulfoxide purchased from Sigma-Aldrich Co. were used as the solvent for preparing polymeric solutions.

Table 1. Polysulfone (PSF) and polyvinylidene fluoride (PVDF) properties (source: datasheet provided by Solvay website).

	PSF (Udel® 3500)	PVDF (Solef® 6008)
Forms	Pellets	Powder
Density (g/cm^3)	1.24	1.75–1.8
Water absorption (24 h, 23 °C)	0.3	<0.04%
Melt temperature (°C)	316–371	170–175
Glass transition temperature (°C)	50	−40
Tensile modulus (GPa)	2.48	1.8–2.5
Tensile strength (MPa)	70.3	30–50
Tensile elongation (%)	50 to 100	20–300%

2.2. Electrospinning

The "lab unit" electrospinning machine by Spinbow company (Bologna, Italy) was used for producing 30 m thick nanofibrous mats. The polymeric solutions of PSF and PVDF were made as follows: (1) PSF solution was prepared by dissolving 23 g of polymer in 90 mL of DMAc and 10 mL of acetone. (2) The second solution was produced by dissolving 15% (*w*/*v*) PVDF powder in a 30:70 (*v*/*v*) of Dimethyl sulfoxide (DMSO) and Acetone. The solutions were poured into two separate syringes and then transferred to the electrospinning machine. The electrospinning parameters are presented in Table 2.

A continuous electrospinning process was conducted for producing pure PSF and PVDF nanofibrous mat, but as the electrospinning machine was not equipped with two separate high-voltage sources and syringe pumps, and due to different feed rates for the two polymers, the process was discontinuous for the mixed (PVDF/PSF) nanofibrous mat: PSF and PVDF nanofibers were electrospun for 1 and 2 min, respectively, until the desired thickness was obtained. SEM pictures of PVDF and PSF nanofibers are shown in Figure 2.

Figure 2. Produced nanofibers: (**a**) PVDF and (**b**) PSF.

Table 2. Electrospinning parameters.

Electrospinning Parameters	PSF (Udel® 3500)	PVDF (Solef® 6008)
Applied voltage (kV)	22	12
Feed rate (mL/h)	1.2	0.6
Collector speed (rpm)	100	100
Needle tip-collector distance (mm)	120	120
Temperature (°C)	25	25

2.3. DCB and ENF Specimens

Double cantilever beam (DCB) and end-notched flexure (ENF) specimens were manufactured and tested under Mode-I and Mode-II fracture loadings, according to ASTM D5528 [40] and guidelines provided by [41], respectively. The samples were manufactured by stacking 14 layers of prepreg woven carbon/epoxy laminates (twill 2/2 240 gsm supplied by Impregnatex Composite Srl) on each other, and the nanofibrous mat and a 15 m thick Teflon layer interleaved between mid-layers. After the lay-up, samples were sealed completely, using a vacuum bag, and transferred to an autoclave to cure: from room temperature to 170 °C (at 1 °C/min), then 1 h at 170 °C, from 170 °C to 190 °C (at 1 °C/min), then 20 min at 190 °C, and finally the oven was shut off and kept closed until complete cooling. Samples were 20 mm wide and 4.2 mm thick, the initial crack length was 59 mm for DCB samples and 40 mm for the ENF ones, and total length was 140 mm (DCB) and 150 mm (ENF). Three samples were produced for each configuration.

2.4. Mode-I Interlaminar Fracture Test

In order to load the samples, aluminum blocks were glued to each side of the samples, as shown in Figure 3. In order to observe the delamination progress by a digital image correlation (DIC) system and measure the crack length (more details in Reference [42]), one side of each sample was coated with a white paint first, and then with a black paint, to obtain a random pattern. The tests were performed in a universal testing machine (Instron 8033), at a constant crosshead speed of 1.5 mm/min. The following expression was used to calculated G_{IC} [40]:

$$G_{IC} = 3F\delta/2Ba, \quad (1)$$

where F is the applied load, δ is the displacement of loading point, B is the width of specimen, and a is the crack length.

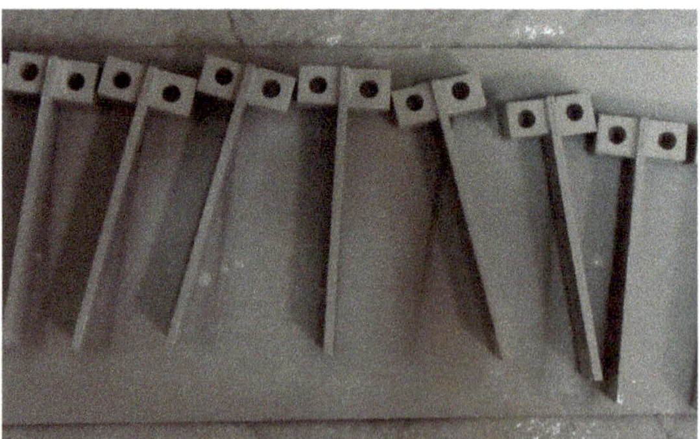

Figure 3. Double cantilever beam (DCB) samples after white painting.

2.5. Mode-II Interlaminar Fracture Test

ENF samples were used to conduct Mode-II fracture tests in a three-point bending load configuration, as shown in Figure 4, at a crosshead speed of 1 mm/min, on the same machined used for Mode-I tests. Span length was 100 mm; therefore, the distance between the crack tip and the loading point was 35 mm. For calculating G_{IIC}, the following formula was applied [41]:

$$G_{IIC} = (4.5a^2 F\delta)/(B(0.25L^3+3a^3)), \quad (2)$$

where a, B, L, F, and δ are the crack length, specimen width, span length, force, and displacement, respectively.

Figure 4. End-notched flexure (ENF) sample under Mode-II test.

3. Results

3.1. Test Results

Figure 5 shows the force-displacement curves for the reference and modified samples under DCB loadings, and Table 3 presents the results. As seen, the PSF and PVDF did not affect the slope of the linear loading phase before crack propagation. An interesting phenomenon is observed while the crack propagates. In the reference laminate, a high number of short rises and falls of the force is registered, unlike the modified laminates, especially the PSF- and PVDF/PSF-modified ones, where a lower number of variations is registered (see the orange ovals in the figures). In nanomodified samples, the force rises after a drop up to about 6N, which is 15% of the maximum load.

The maximum load (Fmax) was 32.7 N for the reference laminate, and it increased 11% and 21% by applying PSF and PVDF nanofibers, respectively; G_{IC} for the non-modified sample is 255 N/m, whereas, for the PVDF, PSF, and PVDF/PSF samples, it is 423, 384 and 454 N/m, respectively. By comparison with the reference, the energy-release rate of PSF- and PVDF-modified laminates was enhanced by 51% and 66%, respectively, and a higher enhancement of 78% was obtained by using the mixed nanofibrous mat.

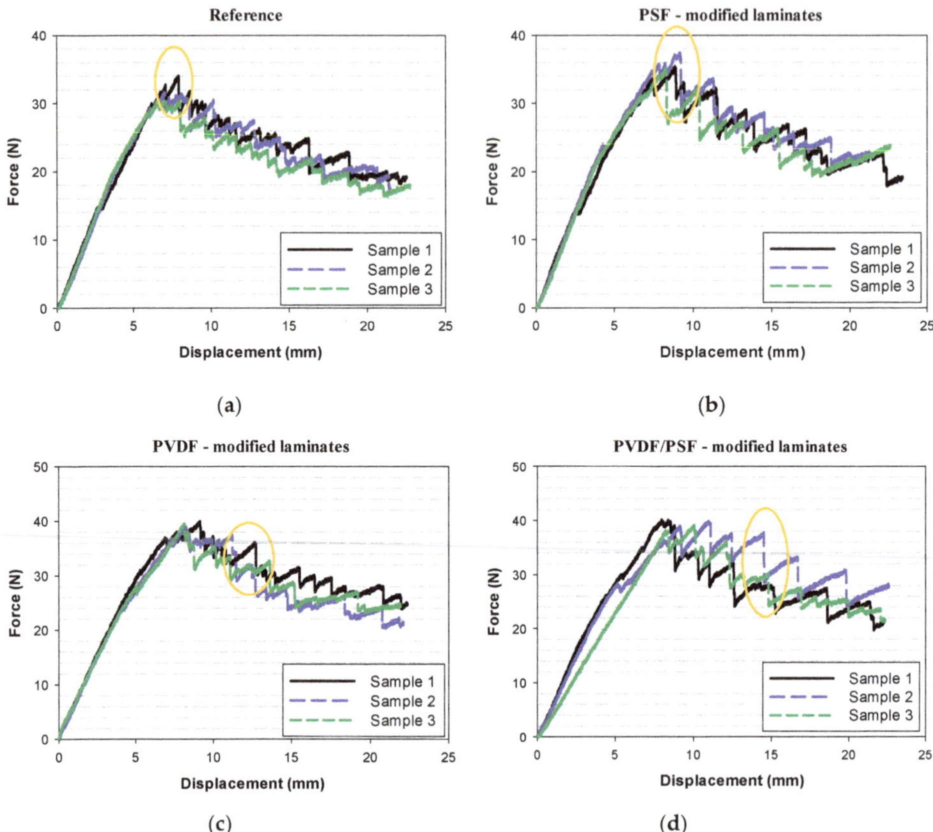

Figure 5. Mode-I fracture test (DCB) outcome for reference (**a**) PSF-only, (**b**) PVDF-only, (**c**) PVDF/PSF (**d**) nanomodified samples.

Table 3. DCB test results.

	Reference	PVDF	PSF	PVDF/PSF
Maximum load (N)	32.7 ± 2	39.6 ± 0.5	36.3 ± 1.5	39.6 ± 0.5
Variation (%)		+21	+11	+21
G_{IC} (N/m)	255 ± 7	423 ± 53	384 ± 9	454 ± 26
Variation (%)		+66	+51	+78

ENF test curves are shown in Figure 6, and the results are presented in Table 4. The behavior of the two types of samples differs at the fracture initiation stage. The crack started to propagate 60–80 N below the Fmax in both control the PSF-modified samples. In this critical point, the slope of force-displacement curve decreases, flagging a crack propagation. In the PVDF-modified laminate, the crack initiation was followed by a force drop, about 40–60 N, in various samples. Then, the force increased again up to the maximum load. The force-displacement curve of the laminates interleaved by PVDF/PSF has some similarities with both of the two other modified samples. In the stage of crack initiation, a very small force drop was observed, and then the load increased about 10–20 N, up to the F_{max} with a lower slope.

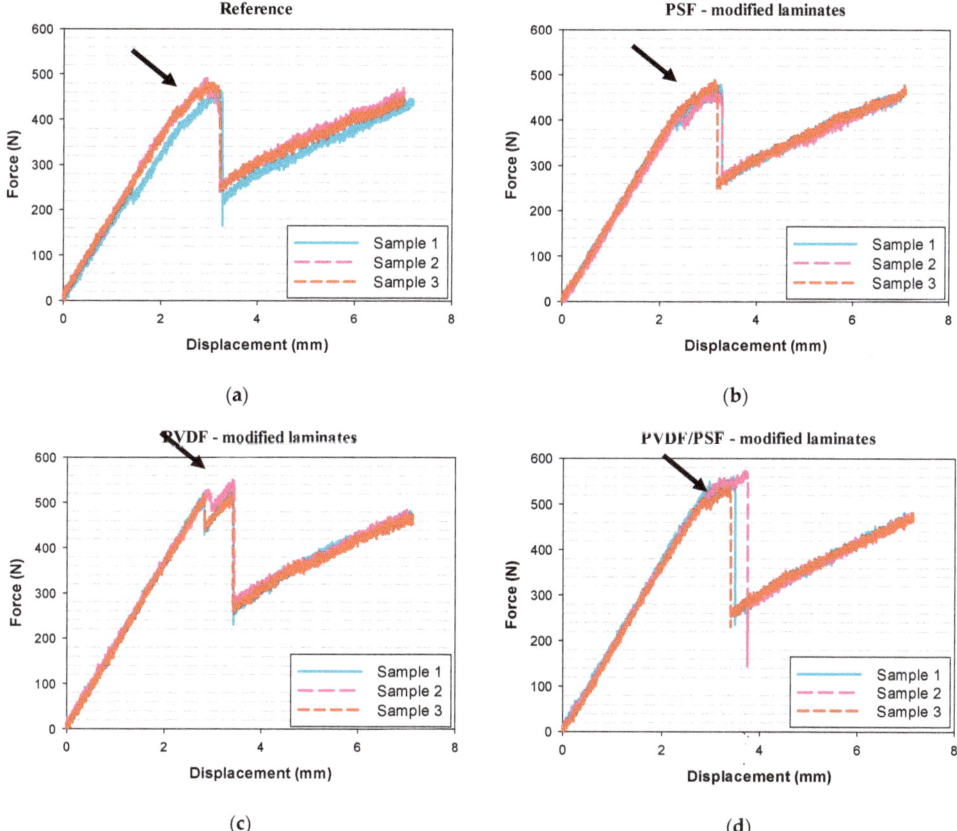

Figure 6. Mode-II fracture test (ENF) outcomes for reference (**a**), PSF-only (**b**), PVDF only (**c**) and PVDF/PSF (**d**) nanomodified samples.

According to Table 4, reference and PSF-modified laminates have similar values of maximum load and G_{IIC}. Therefore, the PSF nanofibers do not show significant effect on toughening the virgin laminate, while its influence in Mode-I loading was positive. On the other hand, PVDF and PVDF/PSF nanofibers increased the G_{IIC} of the laminate by 57% and 75%, respectively. It is interesting to note that, although the influence of PSF nanofibers on G_{IIC} was negligible, its mixture with PVDF had a synergistic effect. A similar phenomenon was observed by Zheng et al. [39]. They used PCL film, nylon nanofibers, and their mixture for toughening carbon/epoxy laminates: According to their results, presented in Table 5, each interlayer individually increased G_{IIC} by 20%, while their mixture almost doubled it.

Table 4. ENF test results.

	Reference	PVDF	PSF	PVDF/PSF
Maximum load (N)	477 ± 13	523 ± 15	478 ± 12	558 ± 16
Variation (%)		+10	/	+17
G_{IIC} (N/m)	182 ± 3	285 ± 2	199 ± 9	318 ± 29
Variation (%)		+57	+9	+75

Table 5. The influence of PCL film, Nylon 66 nanofibrous mat, and their mixture on G_{IIC} [39].

Sample	Reference	Nylon 66 Nanofiber	PCL Film	Nylon 66/PCL
G_{IIC} (N/m)	1420 ± 60	1710 ± 120	1700 ± 90	2820 ± 120
Variation (%)	-	20.4	19.7	98.6

3.2. Toughening Mechanisms

Figure 5 shows the force-displacement curves for the reference and modified samples under DCB loadings, and Table 3 presents the results. As seen, the PSF and PVDF did not affect the slope of the linear loading phase before crack propagation. An interesting phenomenon is observed while the crack propagates. In the reference laminate, a high number of short rises and falls of the force is registered, unlike the modified laminates, especially the PSF- and PVDF/PSF-modified ones, where a lower number of variations is registered (see the orange ovals in the figures). In nanomodified samples, the force rises after a drop up to about 6N, which is 15% of the maximum load.

Figure 7 presents the SEM micrographs of the fractured surfaces of the reference and nanomodified laminates. As seen in Figure 7a, the surface of fractured neat epoxy is smooth, a sign of a brittle type of fracture; instead, for the other samples, images show a different situation and various toughening mechanisms.

Due to high porosity and specific surface area, PSF mats were easily impregnated by the epoxy. On the other hand, owing to PSF's high viscosity and fast curing of resin, the diffusion of PSF in the epoxy was more difficult. Subsequently, the nanofibers were dissolved in the resin. By continuing the curing process, PSF started a phase separation from the epoxy and changed to spherical particles (see Figure 7b). When the crack tip reached these particles, it was restrained and deflected from its path, requiring higher energy to propagate.

4. Discussion

As the curing temperature of composite laminates was higher than the melting point of PVDF (170 °C), the nanofibers melted. As mentioned before, the porous nature of nanofibrous mats caused the epoxy to permeate completely into the PVDF mats before hardening, and, therefore, the PVDF blended with the epoxy by the end of curing process (see Figure 7c). Since PVDF is a thermoplastic polymer, its toughness is higher than thermosets like epoxy; therefore, more energy is required for crack propagation in the blend of PVDF/epoxy. Furthermore, a plastic zone area was detected in front of the crack tip, which could again absorb more energy in comparison with the crack propagated in pure epoxy. Figure 7d illustrates the sample modified by PVDF/PSF nanofibrous mat, showing both the toughening mechanism of each individual nanofiber, i.e., melted PVDF and PSF spherical particles, which both hindered fracture propagation.

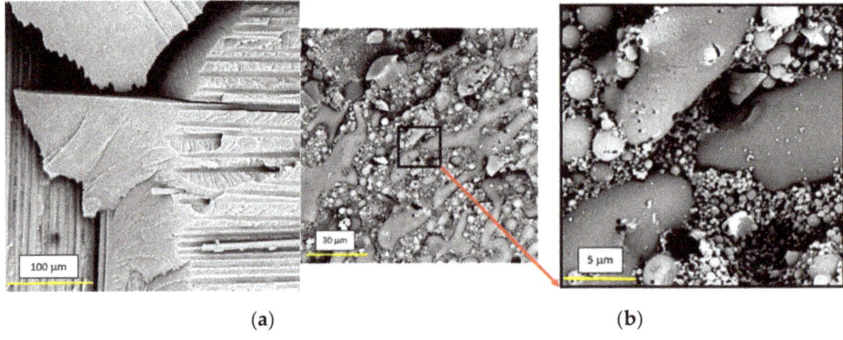

(a) (b)

Figure 7. *Cont.*

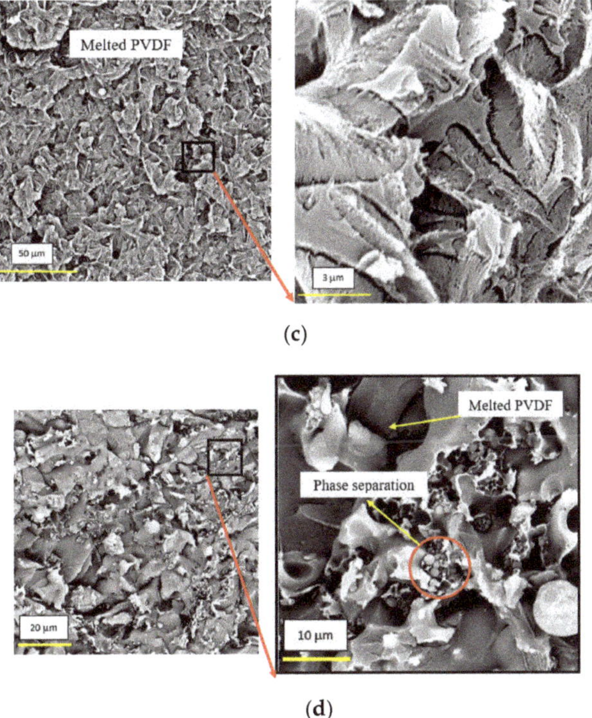

Figure 7. Morphology of fractured surface in (**a**) reference, (**b**) PSF-, (**c**) PVDF-, and (**d**) PVDF/PSF-modified laminates.

5. Conclusions

In this study, two thermoplastic polymers, i.e., PVDF and PSF, were applied individually and at the same time, in form on nanofibers into CFRP, to study their influence on G_{IC} and G_{IIC}. The investigation produced the following results:

1. Interleaving reference laminates with nanofibers made of both polymers lead to increased Mode-I and Mode-II energy-release rates;
2. PVDF nanofibers offer better results than PSF nanofibers: the first improved G_{IC} and G_{IIC} by 57% and 66% against 9% and 51%, respectively, of the latter;
3. Mixed PVDF/PSF nanofibrous mat had a synergistic influence, increasing Mode-I and Mode-II energy-release rate by 75% and 78%, respectively. These results are similar to the results presented in Reference [39], in which PCL film/nylon nanofibers were applied.
4. SEM micrographs showed that PSF started to engage in phase separation from the epoxy and changed to spherical particles during curing process. Hence, when the crack tip reached these particles, it deflected from its main path and absorbed energy.
5. When PVDF nanofibers are interleaved, they mix with epoxy and melt during cure, creating a plastic zone in front of the crack tip, requiring higher energy for it to propagate. The melting is a critical factor; otherwise, the PVDF cannot affectively toughen the laminates.

Author Contributions: Conceptualization, H.S. and R.P.; methodology, H.S., A.Z., and H.H.; investigation, H.S. and R.P.; data curation, T.M.B.; writing—original draft preparation, H.S.; writing—review and editing, R.P.; visualization, T.M.B.; supervision, G.M. All authors have read and agreed to the published version of the manuscript.

Funding: This research received no external funding.

Conflicts of Interest: The authors declare no conflict of interest.

References

1. Wagih, A.; Sebaey, T.A.; Yudhanto, A.; Lubineau, G. Post-impact flexural behavior of carbon-aramid/epoxy hybrid composites. *Compos. Struct.* **2020**, *239*, 112022. [CrossRef]
2. Al-Hajaj, Z.; Sy, B.L.; Bougherara, H.; Zdero, R. Impact properties of a new hybrid composite material made from woven carbon fibres plus flax fibres in an epoxy matrix. *Compos. Struct.* **2019**, *208*, 346–356. [CrossRef]
3. Esmaeeli, M.; Nami, M.R.; Kazemianfar, B. Geometric analysis and constrained optimization of woven z-pinned composites for maximization of elastic properties. *Compos. Struct.* **2019**, *210*, 553–566. [CrossRef]
4. Moallemzadeh, A.R.; Sabet, S.A.R.; Abedini, H.; Saghafi, H. Investigation into high velocity impact response of pre-loaded hybrid nanocomposite structure. *Thin-Walled Struct.* **2019**, *142*, 405–413. [CrossRef]
5. Zhang, D.; Zheng, X.; Wang, Z.; Wu, T.; Sohail, A. Effects of braiding architectures on damage resistance and damage tolerance behaviors of 3D braided composites. *Compos. Struct.* **2020**, *232*, 111565. [CrossRef]
6. Wang, J.; Ma, C.; Chen, G.; Dai, P. Interlaminar fracture toughness and conductivity of carbon fiber/epoxy resin composite laminate modified by carbon black-loaded polypropylene non-woven fabric interleaves. *Compos. Struct.* **2020**, *234*, 111649. [CrossRef]
7. Saghafi, H.; Fotouhi, M.; Minak, G. Improvement of the Impact Properties of Composite Laminates by Means of Nano-Modification of the Matrix—A Review. *Appl. Sci.* **2018**, *8*, 2406. [CrossRef]
8. Palazzetti, R.; Zucchelli, A. Electrospun nanofibers as reinforcement for composite laminates materials—A review. *Compos. Struct.* **2017**, *182*, 711–727. [CrossRef]
9. Saghafi, H.; Brugo, T.; Minak, G.; Zucchelli, A. Improvement the impact damage resistance of composite materials by interleaving Polycaprolactone nanofibers. *Eng. Solid Mech.* **2015**, *3*, 21–26. [CrossRef]
10. Li, P.; Liu, D.; Zhu, B.; Li, B.; Jia, X.; Wang, L.; Li, G.; Yang, X. Synchronous effects of multiscale reinforced and toughened CFRP composites by MWNTs-EP/PSF hybrid nanofibers with preferred orientation. *Compos. Part. A Appl. Sci. Manuf.* **2015**, *68*, 72–80. [CrossRef]
11. Li, G.; Li, P.; Yu, Y.; Jia, X.; Zhang, S.; Yang, X.; Ryu, S.-K. Novel carbon fiber/epoxy composite toughened by electrospun polysulfone nanofibers. *Mater. Lett.* **2008**, *62*, 511–514. [CrossRef]
12. Xuefeng, A.N.; Shuangying, J.I.; Bangming, T.; Zilong, Z.; Xiao-Su, Y. Toughness improvement of carbon laminates by periodic interleaving thin thermoplastic films. *J. Mater. Sci. Lett.* **2002**, *21*, 1763–1785. [CrossRef]
13. Guo, Z.; Li, Z.; Liu, J.; Chen, M.; Zu, Q.; Zhang, Y.; Chen, J. Ex-Situ method for toughening glass/epoxy composites by interlaminar films made of polyetherketone cardo and calcium sulfate whisker. *J. Reinf. Plast. Compos.* **2014**, *33*, 1966–1975. [CrossRef]
14. Adak, N.C.; Chhetri, S.; Kuila, T.; Murmu, N.C.; Samanta, P.; Lee, J.H. Effects of hydrazine reduced graphene oxide on the inter-laminar fracture toughness of woven carbon fiber/epoxy composite. *Compos. Part. B Eng.* **2018**, *149*, 22–30. [CrossRef]
15. Daelemans, L.; van der Heijden, S.; de Baere, I.; Rahier, H.; van Paepegem, W.; de Clerck, K. Improved fatigue delamination behaviour of composite laminates with electrospun thermoplastic nanofibrous interleaves using the Central Cut-Ply method. *Compos. Part. A Appl. Sci. Manuf.* **2017**, *94*, 10–20. [CrossRef]
16. Sarasini, F.; Tirillò, J.; Bavasso, I.; Bracciale, M.P.; Sbardella, F.; Lampani, L.; Cicala, G. Effect of electrospun nanofibres and MWCNTs on the low velocity impact response of carbon fibre laminates. *Compos. Struct.* **2020**, *234*, 111776. [CrossRef]
17. Saghafi, H.; Moallemzadeh, A.R.; Zucchelli, A.; Brugo, T.M.; Minak, G. Shear mode of fracture in composite laminates toughened by polyvinylidene fluoride nanofibers. *Compos. Struct.* **2019**, *227*, 111327. [CrossRef]
18. Magniez, K.; de Lavigne, C.; Fox, B.L. The effects of molecular weight and polymorphism on the fracture and thermo-mechanical properties of a carbon-fibre composite modified by electrospun poly (vinylidene fluoride) membranes. *Polymer* **2010**, *51*, 2585–2596. [CrossRef]
19. Zhang, J.; Yang, T.; Lin, T.; Wang, C.H. Phase morphology of nanofibre interlayers: Critical factor for toughening carbon/epoxy composites. *Compos. Sci. Technol.* **2012**, *72*, 256–262. [CrossRef]
20. Saghafi, H.; Brugo, T.; Minak, G.; Zucchelli, A. The effect of PVDF nanofibers on mode-I fracture toughness of composite materials. *Compos. Part. B Eng.* **2015**, *72*, 213–216. [CrossRef]

21. Kheirkhah Barzoki, P.; Rezadoust, A.M.; Latifi, M.; Saghafi, H. The experimental and numerical study on the effect of PVB nanofiber mat thickness on interlaminar fracture toughness of glass/phenolic composites. *Eng. Fract. Mech.* **2018**, *194*, 145–153. [CrossRef]
22. Barzoki, P.K.; Rezadoust, A.M.; Latifi, M.; Saghafi, H.; Minak, G. Effect of nanofiber diameter and arrangement on fracture toughness of out of autoclave glass/phenolic composites—Experimental and numerical study. *Thin-Walled Struct.* **2019**, *143*, 106251. [CrossRef]
23. Taheri, H.; Oliaei, M.; Ipakchi, H.; Saghafi, H. Toughening phenolic composite laminates by interleaving hybrid pyrolytic carbon/polyvinyl butyral nanomat. *Compos. Part. B Eng.* **2020**, *191*, 107981. [CrossRef]
24. Zheng, N.; Huang, Y.; Liu, H.Y.; Gao, J.; Mai, Y.W. Improvement of interlaminar fracture toughness in carbon fiber/epoxy composites with carbon nanotubes/polysulfone interleaves. *Compos. Sci. Technol.* **2017**, *140*, 8–15. [CrossRef]
25. Saghafi, H.; Minak, G.; Zucchelli, A.; Brugo, T.M.; Heidary, H. Comparing various toughening mechanisms occurred in nanomodified laminates under impact loading. *Compos. Part. B Eng.* **2019**, *174*, 106964. [CrossRef]
26. Saghafi, H.; Ghaffarian, S.R.; Yademellat, H.; Heidary, H. Finding the best sequence in flexible and stiff composite laminates interleaved by nanofibers. *J. Compos. Mater.* **2019**, *53*. [CrossRef]
27. Ahmadloo, E.; Gharehaghaji, A.; Latifi, M.; Saghafi, H.; Mohammadi, N. Effect of PA66 nanofiber yarn on tensile fracture toughness of reinforced epoxy nanocomposite. *Proc. Inst. Mech. Eng. Part. C J. Mech. Eng. Sci.* **2018**, *233*. [CrossRef]
28. Gholizadeh, A.; Najafabadi, M.A.; Saghafi, H.; Mohammadi, R. Considering damage during fracture tests on nanomodified laminates using the acoustic emission method. *Eur. J. Mech. A/Solids* **2018**, *72*, 452–463. [CrossRef]
29. Gholizadeh, A.; Najafabadi, M.A.; Saghafi, H.; Mohammadi, R. Considering damages to open-holed composite laminates modified by nanofibers under the three-point bending test. *Polym. Test.* **2018**, *70*, 363–377. [CrossRef]
30. Yademellat, H.; Nikbakht, A.; Saghafi, H.; Sadighi, M. Experimental and numerical investigation of low velocity impact on electrospun nanofiber modified composite laminates. *Compos. Struct.* **2018**, *200*, 507–514. [CrossRef]
31. Mohammadi, R.; Najafabadi, M.A.; Saghafi, H.; Zarouchas, D. Fracture and fatigue behavior of carbon/epoxy laminates modified by nanofibers. *Compos. Part. A Appl. Sci. Manuf.* **2020**, *137*, 106015. [CrossRef]
32. Mohammadi, R.; Najafabadi, M.A.; Saghafi, H.; Zarouchas, D. Mode-II fatigue response of AS4/8552 carbon /epoxy composite laminates interleaved by electrospun nanofibers. *Thin-Walled Struct.* **2020**, *154*, 106811. [CrossRef]
33. Zhang, H.; Bharti, A.; Li, Z.; Du, S.; Bilotti, E.; Peijs, T. Localized toughening of carbon/epoxy laminates using dissolvable thermoplastic interleaves and electrospun fibres. *Compos. Part. A Appl. Sci. Manuf.* **2015**, *79*, 116–126. [CrossRef]
34. Magniez, K.; Chaffraix, T.; Fox, B. Toughening of a carbon-fibre composite using electrospun poly(hydroxyether of bisphenol A) nanofibrous membranes through inverse phase separation and inter-domain etherification. *Materials* **2011**, *4*, 1967–1984. [CrossRef]
35. Wang, Y.; Wang, Y.; Wan, B.; Han, B.; Cai, G.; Chang, R. Strain and damage self-sensing of basalt fiber reinforced polymer laminates fabricated with carbon nanofibers/epoxy composites under tension. *Compos. Part. A Appl. Sci. Manuf.* **2018**, *113*, 40–52. [CrossRef]
36. Ravindran, A.R.; Ladani, R.B.; Wang, C.H.; Mouritz, A.P. Synergistic mode II delamination toughening of composites using multi-scale carbon-based reinforcements. *Compos. Part. A Appl. Sci. Manuf.* **2019**, *117*, 103–115. [CrossRef]
37. Hsiao, K.T.; Scruggs, A.M.; Brewer, J.S.; Hickman, G.J.S.; McDonald, E.E.; Henderson, K. Effect of carbon nanofiber z-threads on mode-I delamination toughness of carbon fiber reinforced plastic laminates. *Compos. Part. A Appl. Sci. Manuf.* **2016**, *91*, 324–335. [CrossRef]
38. Palazzetti, R.; Zucchelli, A.; Gualandi, C.; Focarete, M.L.; Donati, L.; Minak, G.; Ramakrishna, S. Influence of electrospun Nylon 6,6 nanofibrous mats on the interlaminar properties of Gr–epoxy composite laminates. *Compos. Struct.* **2012**, *94*, 571–579. [CrossRef]

39. Zheng, N.; Liu, H.Y.; Gao, J.; Mai, Y.W. Synergetic improvement of interlaminar fracture energy in carbon fiber/epoxy composites with nylon nanofiber/polycaprolactone blend interleaves. *Compos. Part. B Eng.* **2019**, *171*, 320–328. [CrossRef]
40. ASTM D5528. *Standard Test Method for Mode I Interlaminar Fracture Toughness of Unidirectional Fiber-Reinforced Polymer Matrix Composites*; Annual Book of ASTM Standards; ASTM: West Conshohocken, PA, USA, 2009.
41. European Structural Integrity Society. *Protocol No 2 for Interlaminar Fracture Toughness Testing of Composites: Mode II, Bordeaux*; European Structural Integrity Society: Delft, The Netherlands, 1993.
42. Saghafi, H.; Ghaffarian, S.R.; Brugo, T.M.; Minak, G.; Zucchelli, A.; Saghafi, H.A. The effect of nanofibrous membrane thickness on fracture behaviour of modified composite laminates—A numerical and experimental study. *Compos. Part. B Eng.* **2016**, *101*, 116–123. [CrossRef]

Article

Repair Performance of Self-Healing Microcapsule/Epoxy Resin Insulating Composite to Physical Damage

Youyuan Wang [1], Yudong Li [1,*], Zhanxi Zhang [1], Haisen Zhao [2] and Yanfang Zhang [1]

1 The State Key Laboratory of Power Transmission Equipment & System Security and New Technology, Chongqing University, Chongqing 400044, China; y.wang@cqu.edu.cn (Y.W.); zhx.zhang@cqu.edu.cn (Z.Z.); 20121813107@cqu.edu.cn (Y.Z.)

2 Jinzhong Power Supply Company of State Grid Shanxi Electric Power Company, Shanxi 030600, China; 20134209@cqu.edu.cn

* Correspondence: 20171102031t@cqu.edu.cn; Tel.: +86-1330-837-4882

Received: 3 September 2019; Accepted: 23 September 2019; Published: 1 October 2019

Abstract: Minor physical damage can reduce the insulation performance of epoxy resin, which seriously threatens the reliability of electrical equipment. In this paper, the epoxy resin insulating composite was prepared by a microcapsule system to achieve its self-healing goal. The repair performance to physical damage was analyzed by the tests of scratch, cross-section damage, electric tree, and breakdown strength. The results show that compared with pure epoxy resin, the composite has the obvious self-healing performance. For mechanical damage, the maximum repair rate of physical structure is 100%, and the breakdown strength can be restored to 83% of the original state. For electrical damage, microcapsule can not only attract the electrical tree and inhibit its propagation process, but also repair the tubules of electrical tree effectively. Moreover, the repair rate is fast, which meets the application requirements of epoxy resin insulating material. In addition, the repair behavior is dominated by capillarity and molecular diffusion on the defect surface. Furthermore, the electrical properties of repaired part are greatly affected by the characteristics of damage interface and repair product. In a word, the composite shows better repair performance to physical damage, which is conducive to improving the reliability of electrical insulating materials.

Keywords: self-healing; epoxy resin; microcapsule; insulating composite; breakdown strength; physical damage; electrical tree

1. Introduction

Epoxy resin has been widely used in the field of electrical insulating materials due to its high insulation strength, stable chemical performance, and excellent weatherability [1–4]. However, in the process of manufacture, transportation, and operation, the material can be deteriorated gradually by various factors (such as electrical, thermal, and mechanical factors). The deterioration can lead to the physical damage in material, such as micro-voids and micro-cracks [4–8]. In addition, micro defects can distort the electric field and lead to partial discharge, reducing the insulation performance of material [9–11]. Furthermore, the existing technologies are difficult to detect and repair the damaged parts. And most methods require maintenance after the power outage [11–15]. Therefore, the physical damage has a great impact on the operation of electrical power system. If the insulating material has the self-healing ability, the further deterioration of defects can be prevented in time. So, the electrical and mechanical performances of material can be restored, which can greatly reduce the impact of tiny physical damage on the power system. However, the existing research on self-healing material are mostly related to the mechanical performances of building and coating material, and rarely involve the insulating material [16,17]. Therefore, it is necessary to develop the self-healing epoxy resin insulating composite. It can prolong the service life of insulating material fundamentally and improve the reliability of electrical power system.

Recently, the self-healing technologies are mainly divided into two types: intrinsic and extrinsic [17–21]. The repair behavior of intrinsic material is mainly realized by the reversible chemical reaction or the diffusion of macromolecules [18,19]. However, the intrinsic method has higher requirements on the properties of material, and its application range has some limitations. Compared with the intrinsic material, the extrinsic material has better weatherability, wider application range and more manufacturing process [20,21]. Moreover, among them, the microcapsule system has higher stability, better repair rate and less damage to the matrix material structure [16,17,21]. Considering the operation environment and repair requirements of insulating material, the microcapsule system was selected as the research object in this paper. For microcapsule/epoxy resin composite, the self-healing characteristic of electrical performances have not been involved yet, and the repair behavior is still lack of systematic mechanism [20–22].

At present, there are few reports about self-healing insulation material. In the aspect of simulation, the static stress distribution of the microcapsule used for epoxy resin insulating material was simulated [2]. The results confirmed that the internal stress of insulating material can lead to the rupture of the microcapsule, thereby completing the repairing behavior. However, it does not involve the repairing ability of actual insulating material. In terms of experiment, the ability of microcapsule to repair electrical tree damage in epoxy resin was studied [23]. The results showed that the microcapsules can delay the development of electrical trees. But the change rules and mechanism of repair performance are still insufficient. In addition, the migration behavior of superparamagnetic nanoparticles was used to repair the electrical tree damage in polymers [24]. However, this method requires the stimulation of an external magnetic field, which belongs to the inductive repair system. There is still a certain gap with the active self-healing target of insulating material. Therefore, the research on the self-healing epoxy resin insulating material remains to be deepened.

In this paper, based on the preparation of microcapsule/epoxy resin composite, its self-healing performance to physical damage was analyzed by the tests of scratch, cross-section, electric tree, and breakdown. Furthermore, the repair mechanism was analyzed by the capillary theory and molecular diffusion model on the defect surface. In addition, the change of alternating current (AC) breakdown strength of epoxy resin composite was studied innovatively. Combined with the reaction mechanism and the interface electric field model, the changes of composite electrical insulation performance were explained. The goal of this work is to explore the repair ability of insulating material to physical damage and prolong the service life of insulation material fundamentally. Furthermore, the work in this paper can lay an experimental foundation for the development of self-healing insulation material.

2. Materials and Methods

2.1. Materials and Preparation

In the existing research, the microcapsule system of urea-formaldehyde (UF) resin coated epoxy resin is mostly selected to achieve the self-healing behavior in epoxy resin matrix [2,17,20]. However, previous studies did not consider the application requirements of epoxy resin insulating material. Although the compatibility of the epoxy repairing agent and the epoxy matrix is good, the curing rate of epoxy repairing agent is slow. It is not suitable for the timely requirement of repair behavior in insulating material. Especially, in high electric field, the epoxy repairing agent cannot suppress the partial discharge in time. The reaction rate of dicyclopentadiene (DCPD) is faster. And the dielectric constant of its reaction product (i.e., polydicyclopentadiene (PDCPD)) are close to those of epoxy resin, which can effectively homogenize the local high field of the defect, thus reducing the impact of structural damage [23]. Therefore, the UF/DCPD microcapsule was selected in this paper.

2.1.1. Preparation of Microcapsule

In this paper, the microcapsule was prepared by the "two-step" in-situ polymerization. The two-step method is to first obtain the UF prepolymer. And then the prepolymer undergoes the polycondensation reaction in the core emulsion to form UF shell. Therefore, the two-step reaction process is stable and controllable, and the performance of the products are superior. The raw materials used are all analytical reagent (AR) level.

(1) UF prepolymer

First, the urea was dissolved in deionized water, then formaldehyde solution was added, and the pH was adjusted to 8.0~9.0. The reaction time was 1 h. at 70 °C. After the solution was cooled to room temperature, the UF prepolymer was obtained. The mass ratio of urea to formaldehyde was about 1:2.3 to ensure the sufficient content of dimethylol urea in prepolymer, which can enhance the net structure of UF product.

(2) Microcapsule

The DCPD emulsion was obtained by mixing sodium dodecyl benzene sulfonate (SDBS) emulsifier, melted DCPD and deionized water at 400 rpm for 30 min. In order to obtain the good dispersion and stability of emulsion, the dosage of SDBS was about 5% of the DCPD mass. Subsequently, the UF prepolymer, the UF curing agent ammonium chloride and the UF water resistant modifier resorcinol were added in the emulsion. The mass ratio of UF prepolymer to DCPD was about 2:1 to ensure the encapsulation effect of microcapsule. The dosage of ammonium chloride and resorcinol were about 1.5% and 2.5% of the prepolymer respectively to ensure the curing and modification effect of wall material. After that, the emulsion pH was slowly adjusted to about 3.0. The acidification time was controlled about 30 min to obtain the intact sphericity of microcapsule. Finally, the microcapsules were obtained by reaction at 60 °C for 3 h.

2.1.2. Preparation of Self-Healing Epoxy Resin Insulating Composite

The temperature factor has a great influence on the curing reaction rate of epoxy resin and material properties. Thus, the room temperature curing system of epoxy resin-low molecular weight polyamide was selected to facilitate the operation of doping microcapsule and ensure the basic properties of insulating material. The raw materials used are all AR level.

First, 80 parts of epoxy resin E 51 was diluted with 20 parts of epoxypropane butyl ether 660 as the matrix. Then 60 parts of room temperature curing agent polyamide, 3 parts of curing accelerator tri (dimethylamine methyl) phenol (DMP-30) and 10 parts of toughening agent dibutyl phthalate were added in matrix to obtain the epoxy room temperature curing system. After that, the microcapsule and the catalyst for core material (i.e., repairing agent) were mixed into the epoxy room temperature curing system to obtain the microcapsule/epoxy resin composite system. In addition, Grubbs' second-generation catalyst with better stability and catalytic efficiency was selected to improve the effect of self-healing. And the dosage of catalyst was 10% of the microcapsules mass.

The composite system was mixed for 30 min at room temperature to ensure uniform dispersion. Then the bubbles in the system were removed by ultrasonic oscillation for 1 h and vacuum treatment for 20 min. Finally, it was cured for 4 days in the room temperature (25 ± 1 °C).

According to the relevant literature [16,25] and my previous work (as shown in Table 1), the important basic properties (such as thermal stability, Young's modulus and tensile strength) of composite with about 1 wt. % microcapsule are more in line with the application requirements of epoxy resin insulating material.

Table 1. Basic performances of composite with different concentrations of microcapsule.

Basic Performances of Composite	Concentrations of Microcapsule in Composite				
	0 wt. %	0.5 wt. %	1 wt. %	5 wt. %	10 wt. %
Repair efficiency to scratch about 25 μm wide (%)	0	≥50	≥70	≥73	≥76
Thermal decomposition temperature (°C)	396.8 ± 1.5	399.1 ± 1.9	402.7 ± 2.3	400.3 ± 2.1	391.0 ± 3.9
Young's modulus (GPa)	2.67 ± 0.03	2.63 ± 0.13	2.50 ± 0.21	2.34 ± 0.17	2.11 ± 0.33
Tensile strength (MPa)	51.6 ± 2.7	50.0 ± 1.9	46.1 ± 3.1	40.3 ± 2.9	32.6 ± 3.7

As the insulating material for industrial products, the thermal and mechanical properties of epoxy resin are crucial for its application. Moreover, the composite with 1 wt. % microcapsule has the best thermal stability and better mechanical properties on the premise of higher repair efficiency. Preliminary analysis shows that the effect of microcapsule on the basic properties of polyethylene is mainly related to the introduction of interface and impurities by microcapsule.

On the one hand, the local state formed by the interface structure can anchor the macromolecular chains in the matrix material, improving the stability of matrix structure [25]. And the stable internal structure can hinder the invasion of external factors (such as thermal factor). Thus, appropriate number of interfaces can improve the properties of material (such as thermal stability). On the other hand, the excessive impurities and interface defects can increase the defect structure in the matrix material. Furthermore, the dispersion of microcapsules will be deteriorated with the increase of microcapsule concentration, affecting the properties of matrix material. Thus, higher concentration of microcapsules can reduce the properties of epoxy resin (such as mechanical properties and thermal stability). Therefore, the performances of composite are affected by the concentration of microcapsule.

Overall, when the dosage of microcapsule was about 1% of the epoxy matrix mass, the comprehensive properties of composite are better. Thus, the more appropriate concentration (1 wt. %) was selected to study the repair performance of composite to physical damage and exclude the influence of other secondary factors.

2.2. Methods

In this paper, the repair performance of composite to physical damage was verified from mechanical damage and electrical damage. Moreover, the effect of microcapsule on the tree discharge and insulation strength of epoxy resin were explored.

2.2.1. Mechanical Damage Test

The scratch damage was used to simulate the mechanical damage in epoxy resin insulating material to verify the repair behavior to physical damage. The scratches were carried out with China Tianchuang QHZ scratch tester. The morphology of scratches was observed by China Guanggu SGO-PH80 optical microscope (OM). According to the optimum polymerization temperature (about 45 °C) of repairing agent (DCPD), the curing temperature (above 60 °C) of reaction product (PDCPD) and the normal operating temperature (less than 155 °C) of epoxy resin insulating material, the temperatures of 60 °C were selected in this paper. The heating time was 30 min and 60 min, respectively.

Moreover, the alternating current (AC) breakdown strength of samples was carried out to verify the self-healing effect to mechanical damage in electrical insulation performance. The breakdown strength was tested on the platform constructed by Ningxia High Voltage Electronic Instrument Company JNC801 transformer. The cylinder-plate electrodes manufactured according to GB/T 1048-2006 standard were used as the electrodes, which are made of brass. In addition, the electrodes and samples were immersed in 25# mineral insulating oil to prevent the surface discharge. The breakdown test for each type of samples was repeated 20 times.

Furthermore, the cross-section test was carried out to further observe the reaction process of repairing agent. And the Czech Tescan MIRA3 scanning electron microscope (SEM) was used to make an intuitive comparison between the pure sample and composite.

2.2.2. Electrical Damage Test

The self-healing ability of composite to electrical damage was verified by electrical tree. The needle-plate electrode system was used to initiate electrical tree. The needle electrode was inserted into the sample during the preparation process, while the curvature radius of the needle tip was 5 μm. The plate electrode was contacted the sample through conductive adhesive, and the distance between electrodes was 3 mm. Moreover, the sample was immersed in transformer oil to prevent the surface flashover. In the experiment, using the pressure platform in Section 2.2.1, the AC voltage of 10 kV/50 Hz was applied for 60 min at room temperature. The morphological characteristic of electrical tree was observed by OM.

3. Results and Discussion

3.1. Repair Performance of Mechanical Damage

3.1.1. Structural Change of Scratch

Figure 1 shows the OM results of pure epoxy resin and its composite. Compared with the pure epoxy resin, there are obvious microcapsules in composite. And the microcapsules have excellent dispersion in the epoxy resin matrix without obvious rupture. The excellent dispersion and intact morphology of microcapsule in the matrix can ensure the better repair effect of damage based on the less influence on the basic properties of matrix material.

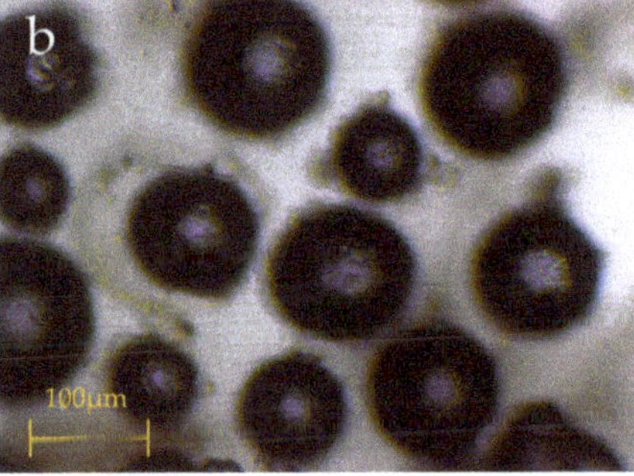

Figure 1. Morphology of epoxy resin samples. (**a**) Pure epoxy resin. (**b**) Microcapsule/epoxy resin composite.

Figure 2 is the OM results of the scratch test. The composite without catalyst was added as the contrast to verify the repair effect to scratch damage more intuitively. The scratch could cause physical structural damage to epoxy resin insulating material. And the scratch width in pure epoxy resin had no obvious change after heating. The results indicate that the original epoxy resin insulating material does not have any self-healing ability to physical damage. Moreover, the influence of heating factor on the physical damage (such as scratch) in epoxy resin material is very small, which can be ignored.

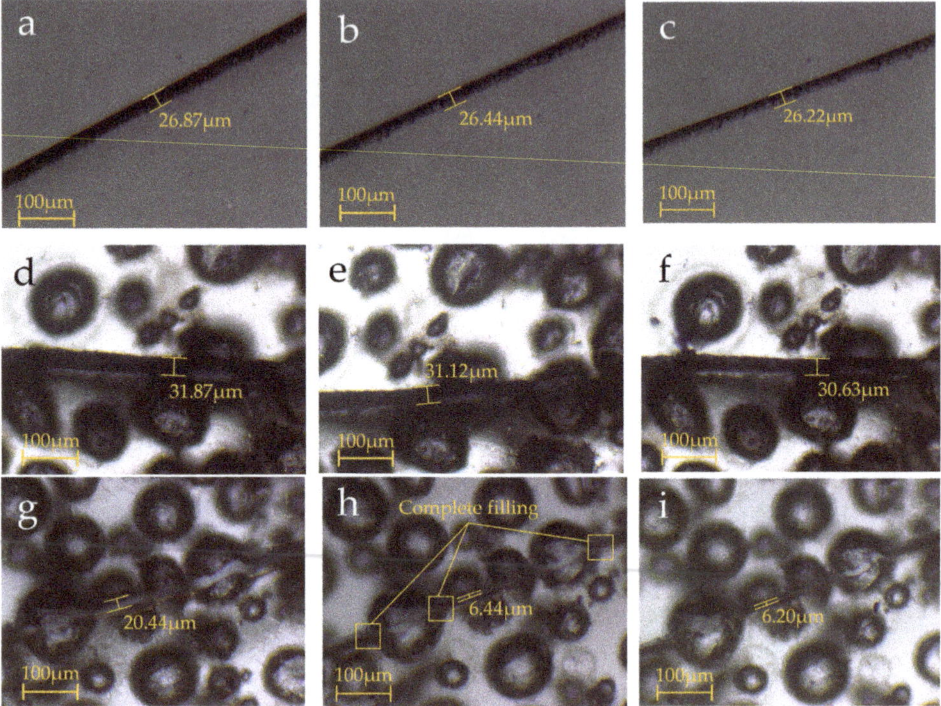

Figure 2. Morphology of the scratch: (**a**) Unheated pure LDPE epoxy resin; (**b**) Heated pure epoxy resin for 30 min; (**c**) Heated pure epoxy resin for 60 min; (**d**) Unheated composite without catalyst; (**e**) Heated composite without catalyst for 30 min; (**f**) Heated composite without catalyst for 60 min; (**g**) Unheated composite; (**h**) Heated composite for 30 min; and (**i**) Heated composite for 60 min.

The scratch can destroy the microcapsules in the sample, resulting in the rupture of wall structure. And the heating condition would cause the core material to flow out. For the composite without catalyst, the unreacted DCPD was volatilized by heating. So, the scratch of this composite had no repair effect, which is similar to the pure epoxy resin. When the composite contained catalyst, its repair behavior is shown in Figure 3. The melted repairing agent (DCPD) flowed to the damaged area, contacted with the catalyst and reacted to form solid repair product (PDPCD) at the defect area. In other words, the microcapsule can fill the damage structure. Therefore, the scratch structure was reduced dramatically. If the scratch width is taken as the criterion of the repair efficiency, the self-healing rate is 70%~100%.

Furthermore, infrared results (as shown in Figure 4) show that the filler in scratch is really the PDCPD (3050 cm^{-1} and 3003 cm^{-1} indicate = C–H, 2926 cm^{-1} and 2851 cm^{-1} indicate $–CH_2–$, 1701 cm^{-1} and 1633 cm^{-1} indicate C=C) [26]. In addition, the carbon-carbon double bonds are retained after the reaction, which makes the stiffness and toughness of product reach an excellent balance. Thus, the good repair effect of the damaged part is ensured.

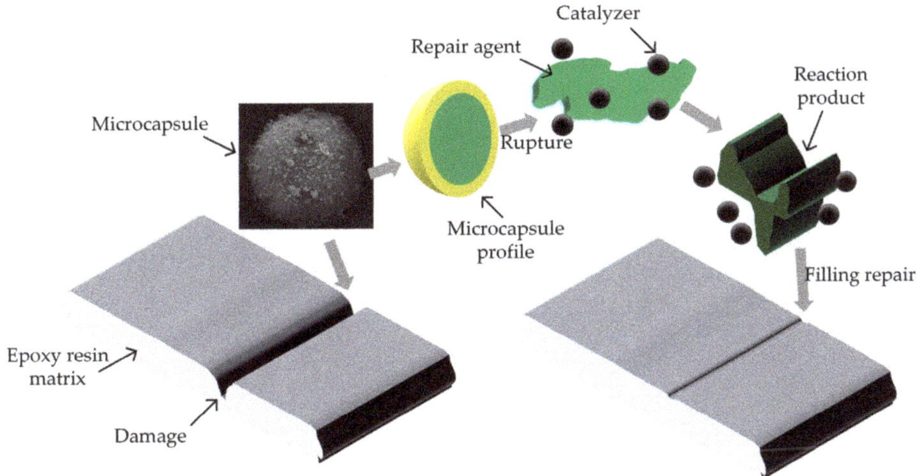

Figure 3. Repair behavior of composite.

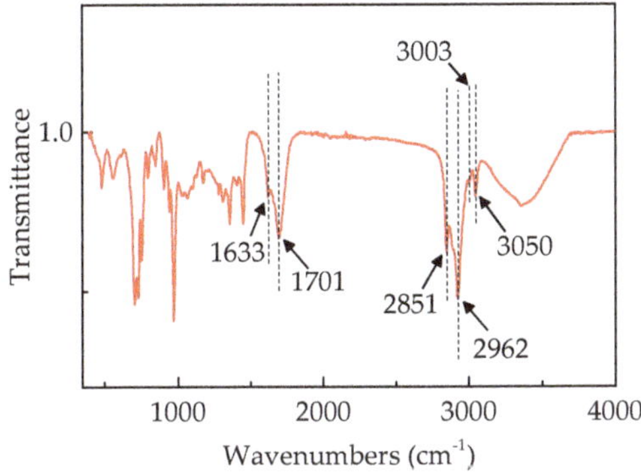

Figure 4. Infrared spectroscopy analysis of the filler in scratch.

The repair effect around the region of ruptured microcapsule is more obvious, and the scratch is almost completely repaired. The main reason is that the damage region near the microcapsule would contact a large number of core material preferentially. Moreover, the reaction rate of core material is relatively fast, leading to the repairing agent having already begun to react in the diffusion process. Furthermore, the comparison between Figure 2h,i shows that the reaction process of core material is concentrated in the first 30 min. It proves that the repair reaction speed is fast, which can better meet the timely requirement of damage treatment in insulating material. Thus, the repair ability of microcapsule can effectively and timely prevent the further deterioration of insulating material.

In short, the obvious self-healing ability of composite to physical damage can be proved by the comparison of three groups of scratch experiments. The repair performance of insulating material to physical defects is realized, thus ensuring the reliable operation of electrical equipment.

3.1.2. Structural Change of Cross-Section

The SEM results of cross-section are showed in Figure 5. From Section 3.1.1, the repair behavior was concentrated in the first 30 min. So, the samples of cross-section were heated for 30 min. Compared with the pure epoxy resin, there are obvious microcapsules in the composite. And the dispersion of microcapsules is great, which is similar to the OM results in Figure 1. Moreover, the microcapsule has obvious core-wall structure and it can be damaged in the cross-section operation (as shown in Figure 5c). It indicates that the wall structure of microcapsule has the rupture response to the stress change of insulating material.

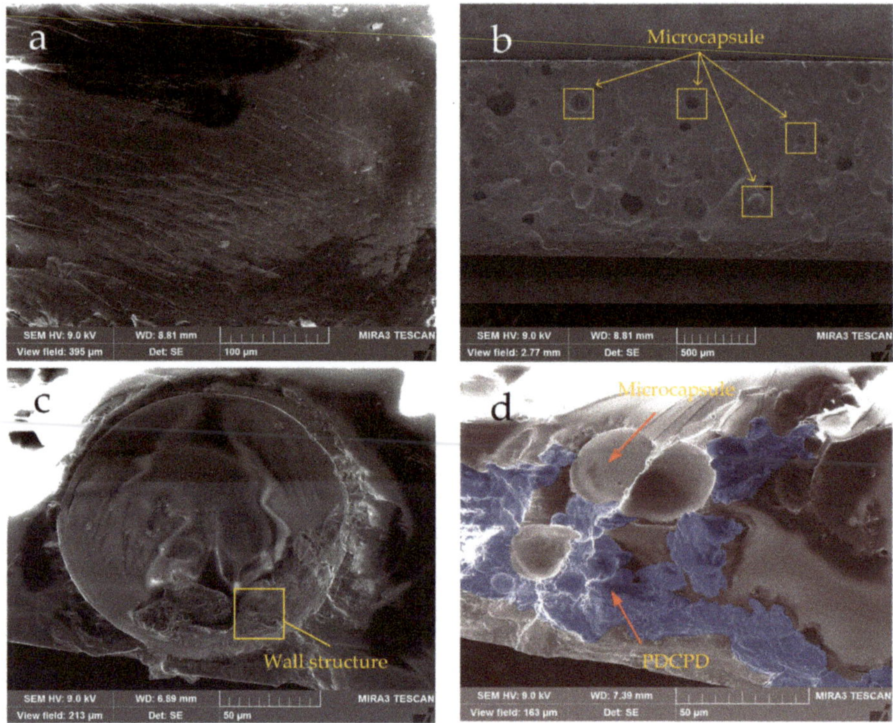

Figure 5. Cross-section characteristics of epoxy resin. (**a**) Pure epoxy resin; (**b**) and (**c**) Composites before reaction; and (**d**) Repair effect.

The reaction effect of the repairing agent is shown in Figure 5d. When the microcapsule was damaged, the core material flowed out and reacted in the damage area. And the PDCPD product appeared as the irregular flocculent material at the cross-section. The flow of DCPD due to the gravity influence is the main reason for the concentration of repair product at the section bottom. This phenomenon also proves that he fluidity of core material is well, which meets the design requirement of repair behavior.

The repair mechanism can be analyzed by combining the capillary flow theory and the molecular diffusion model on defect surface [27]. With the appearance of the minor damage in matrix, the pressure difference would be formed on both sides of the defect. And due to the DCPD is the Newtonian liquid between 32.5~172 °C, the repairing agent would continuously diffuse into the defect depth under the capillary effect. Meanwhile, the core material covers the defect interface to form the surface or body containing the repairing agent. And then the repairing agent molecules attract and react with each other. Although the molecular weight of DCPD is small, its molecular space structure is complex. So, the arrangement of molecule in the reaction process is irregular, resulting in the irregular morphology of repair product. Therefore, the core material of microcapsule can achieve the better repair effect to the physical damage, but the regularity of product is still needed to be improved.

All in all, the experimental results intuitively demonstrate that the doping effect of microcapsule in epoxy resin insulating material is excellent. The wall structure of microcapsule has timely response to the stress change in matrix. And the fluidity and reactivity of core material is great. Therefore, it is proved that the epoxy resin insulating composite based on microcapsule system can repair the damaged parts quickly and automatically.

3.1.3. Change of Insulation Strength to Scratch Damage

Breakdown strength is one of the most important macroscopic performances of insulating material, which can be affected by damage defects. Moreover, the scratch defect can equivalent to the structural damage. Therefore, alternating current (AC) breakdown strength of damaged samples is measured. From Section 3.1.1, the damaged samples were heated for 30 min.

This experiment can reflect the influence of physical damage on the epoxy resin insulating material and the repair effect of composite on its electrical properties. Due to the breakdown data usually have great dispersion, the Weibull distribution (as shown in Formula (1)) is used to reduce the error [28].

$$F(U) = 1 - \exp\left[-\left(\frac{U}{\alpha}\right)^{\beta}\right], \tag{1}$$

where, U is the breakdown voltage value; α is the scale parameter, indicating the breakdown voltage value when the material failure probability is 63.2%; and β is the shape parameter, indicating the dispersion of data.

The AC breakdown strength is shown in Figure 6. The breakdown strengths of epoxy resin before and after doping the microcapsule are maintained at about 45 kV/mm, which indicates that the 1 wt. % microcapsule has little effect on the insulation strength of epoxy resin. On the one hand, the impurities and interface defects introduced by microcapsule will destroy the original continuous structure of matrix material. It can make the internal electric field distribution uneven, thus reducing the electrical insulation performance of matrix [29]. On the other hand, the microcapsule distributes evenly in epoxy resin, which can reduce the influence of impurities on the internal structure of matrix. In addition, the local state can be formed by the interface between microcapsule and epoxy matrix, resulting in many charge traps. The migration of carrier can be affected by charge traps. In other words, the microcapsule can shorten the average free path of carrier, which improves the insulation performance of matrix material [1]. Therefore, the breakdown strength of composite is not decreased significantly, which can meet the application requirements of epoxy resin insulating material.

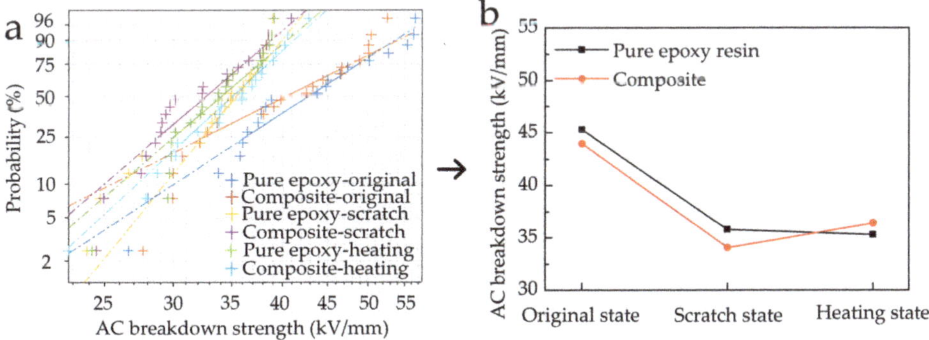

Figure 6. (**a**) Weibull probability. (**b**) Alternating current AC breakdown strength of epoxy resin.

After the scratch treatment, the breakdown strengths of the pure sample and its composite are all reduced by about 10 kV/mm. Moreover, the breakdown point occurs at the scratch site. It is proved that the physical defect (such as a scratch) can directly affect the electrical insulation performance of insulating material, threatening the safe operation of electrical equipment. And the similar decrease amplitude of breakdown strength proves that the 1 wt. % microcapsule has little effect on the original properties of epoxy resin.

There are three main reasons for the influence of scratch on breakdown strength: (1) According to the theory of electrical-mechanical aging, the micro-defect in the material can distort the electric field and form a high field strength area at the damage site [11]. And the electric field at the scratch was more concentrated, so the breakdown strength was reduced. (2) Physical damage would directly reduce the thickness in the direction of breakdown. At the same external electric field, the field strength at the scratch was greater. Therefore, the breakdown would occur at the scratch. (3) The physical damage can be equivalent to the filling of another dielectric. In this experiment, the oil used in the test replaced the epoxy resin in the damaged parts. And the insulation strength of oil is lower than that of epoxy resin. Considering the interface charge effect, more space charge would be accumulated at the interface between oil and epoxy resin [29], which exacerbates the electric field distortion and caused the breakdown phenomenon.

After heating, the breakdown strength of pure epoxy resin is fluctuated slightly, and there is no obvious recovery. Therefore, the physical damage (such as a scratch) has a greater impact on the electrical breakdown characteristics of insulating material. And the degradation of insulation property is permanent and cannot be recovered by itself. Compared with the pure sample, the insulation strength of the composite obviously rises after heating. Moreover, the breakdown strength of composite can be restored to about 83% of its initial state.

The microcapsule can not only repair the physical structure of damage site, but also eliminate the impact of damage on the electrical insulation property of the epoxy resin. On the one hand, the thickness of the defect in the breakdown direction is increased by the filling ability of the microcapsule, improving its microstructure. In other words, the repair ability of the microcapsule can inhibit the distortion and concentration of the electric field at the damage site. On the other hand, the dielectric constant of the repair product (PDCPD) is about 2.78 which is closed to that of epoxy resin (about 3). Therefore, the repair product can effectively uniform the local high electric field and reduce the electrical-mechanical stress of the damage site [11]. Thus, the development of electrical breakdown can be effectively inhibited. In addition, the local state can be introduced by the interface area between the repair product and matrix martial. The discharge energy initially concentrated on the defect can be dispersed by the charge transport of the lower barrier around the local state [1]. Therefore, the electrical insulation property of damaged composite can be restored after repair.

Based on the self-healing mechanism, the incomplete repair phenomenon of breakdown strength in composite is explained in this paper. On the one hand, the reaction rate of repairing agent is too fast and the scope of scratch damage is relatively large. So, the reaction process of the repairing agent may have been completed when the damaged part is not completely covered by the repairing agent. In other words, a part of the physical damage gap is still retained, which is similar to the repair result in Figure 2h. Moreover, as shown in Figure 5d (i.e., the reaction result of cross-section), the surface of the product is uneven and there is a physical difference between the product and the matrix. Thus, there are still some physical defects in the damaged area after repair. So, the insulation performance of the material cannot be restored to its initial state. On the other hand, although the electrical properties of PDCPD and epoxy resin are similar, there are still some differences in the material properties. The density of PDCPD (about 1.03 g/cm^3) is less than that of epoxy resin (about 2 g/cm^3), so the local low-density area will be formed in the repair site [5]. Thus, the disorder of structure and the density of charge trap are increased, reducing the electrical insulation property. Therefore, the breakdown strength of the damaged composite can be raised after the repair, but it still cannot be restored to the initial state.

In general, the experiment results prove that the composite can do repair its electrical insulation property. Therefore, the decrease of breakdown strength caused by physical damage in insulating material can be alleviated, improving the reliability of the power system. At present, the research on self-healing insulating material is still in the exploratory stage. Although a small part of insulation strength is sacrificed, composite can quickly restore its insulation performance after damage. Thus, the composite has good research value.

3.2. Repair Performance of Electrical Damage

In the actual operation process, the electrical tree is the main factor leading to the insulation failure of insulating material [30]. The electrical tree can cause the irreversible structural damage of micron and above, thus directly threatening the safe operation of power system [30,31]. Therefore, the self-healing performance to electrical tree can greatly improve the service life of insulating material. However, there are still few reports about it. This paper mainly studies the propagation stage of electrical tree in epoxy resin. The results are shown in Figure 7. From Section 3.1.1, the heating condition is 60 °C/30 min. This paper observes the panorama of the electrical tree by the mosaic of multiple graphs to ensure the clarity.

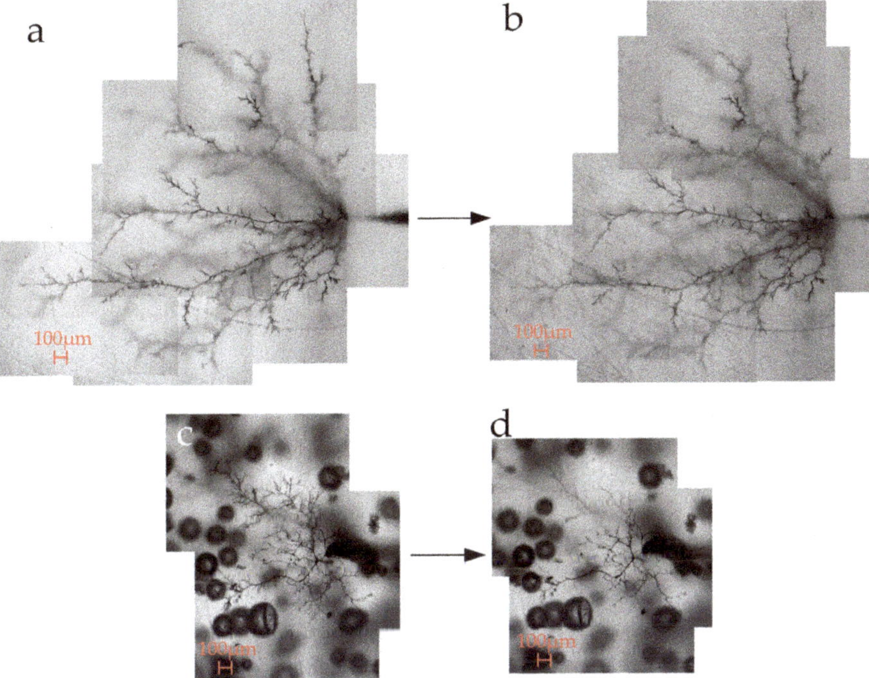

Figure 7. Morphology of the electrical tree. (**a**) Unheated pure epoxy resin. (**b**) Heated pure epoxy resin. (**c**) Unheated composite. (**d**) Heated composite.

Compared with pure epoxy resin, the size of the electrical tree in composite is decreased significantly, which is similar to the result in [23]. In addition, the electrical tree in composite tends to develop into a microcapsule and branches near the microcapsule. It can be concluded that the local electric field can be affected by the microcapsule, which attracts the propagation of the electrical tree. When the electric tree develops into the microcapsule, it can break the wall structure of the microcapsule and consume some energy. The residual energy will turn into a new branch, thus the propagation of the electric tree is inhibited. On the other hand, the continuity of the original structure in epoxy resin can be destroyed by microcapsule, thus hindering the propagation of the electrical tree. Therefore, the electrical tree in the composite tends to develop towards the microcapsule area, and its overall size is decreased.

After heating, the morphology of electrical tree in pure epoxy resin has no obvious change. In composite, except for the main branch of electric tree, the other branches disappeared after heating. Moreover, the width of the main branch in the composite is decreased after heating. Thus, the composite exhibits the obvious self-healing ability to electrical damage. The repair mechanism is similar to that described in Section 3.1, the wall structure of the microcapsule can be broken by the electrical tree, while the core material (repairing agent) can flow into the tubules of the electrical tree through the action of capillary and molecular diffusion. Thus, the partial tubules of the electric tree are filled with repaired product. In addition, due to the dielectric constant of the reaction product being close to that of the epoxy resin, the compatibility between the PDCPD and the matrix is good. Therefore, the local high electric field is weakened, which can further inhibit the development of the electric tree.

Due to the influence of the voltage condition in this paper, the size of the electrical tree is relatively large. Thus, the repairing agent cannot completely repair all the tubules of the electrical tree. However, the temperature and electric field coexisted in the actual operation of the epoxy resin insulating material. And the propagation time of the actual electrical tree is longer [23,30]. Therefore, the repairing agent in the microcapsule has a good opportunity to repair the electrical tree in its early stage. In other words, on the premise of reasonable control of the microcapsule concentration, epoxy resin insulating composite is fully capable of repairing the electrical damage.

The damage to the electrical tree has been regarded as an irreversible permanent defect [24]. In this paper, the microcapsule can not only inhibit the propagation of electrical tree, but also repair the tubules of electrical tree effectively. It has good application value in reducing the harm of electrical damage to insulating material.

4. Conclusions

In this paper, the preparation method and self-healing performance of the microcapsule/epoxy resin insulating composite were explored exploitively and creatively. Moreover, the repair performance to physical damage and the change rule of electrical insulation performance are emphatically studied. The application advantage of self-healing insulating composite is proved by linking the micro-changes with the macro-properties. The main conclusions are as follows:

(1) For the mechanical damage, the microcapsule/epoxy resin composite exhibit excellent self-healing ability. The repair efficiency of scratch width is 70%~100%, and the repair rate is relatively fast. Moreover, the core material of the microcapsule has great fluidity and reactivity. In addition, the repair process is mainly affected by the action of the capillary and molecular diffusion model on the defect surface.

(2) For the electrical damage, the repair effect of composite to the electrical tree is obvious. The microcapsule can not only attract the electrical tree and inhibit its propagation, but also repair the tubules of electrical tree effectively. It has good application value in reducing the harm of electrical damage to insulating material.

(3) Physical damage can reduce the insulation strength of material, thus threatening the stability of the power system. The composite can not only fill the structural defect caused by electrical damage (such as an electrical tree) and mechanical damage (such as a scratch), but also homogenize the local high electric field in the defect area, restoring its insulation strength. The restoration effect of electrical insulation performance is mainly affected by the reaction effect of repairing agent and the interface characteristic between the repair product and the matrix material.

In addition, the microcapsules are uniformly dispersed in epoxy resin insulating material with great morphology. It is conducive to repairing the damaged parts effectively on the premise of ensuring the basic properties of the material (such as thermal and mechanical properties).

Author Contributions: Conceptualization, Y.W. and Y.L.; Methodology, Y.L.; Validation, Y.W., Y.L., and Z.Z.; Data curation, Y.Z., Z.Z., and H.Z.; Formal analysis, Y.W. and Y.L.; Writing—Original draft preparation, Y.L.; Writing—Review and editing, Y.W.; Visualization, Y.L. and Y.W.; Project administration, Y.W.

Funding: This research was funded by National Natural Science Foundation of China (51777018), Fundamental Research Funds for the Central Universities (2019CDXYDQ0010) and the Science and Technology Project of SGCC under Grant (SGTYHT/14-JS-188).

Conflicts of Interest: The authors declare no conflict of interest.

References

1. Liu, C.; Zheng, X.; Bie, C. Research of Preparation and Non-Linear Conductivity Modification of Doped ZnO/Epoxide Resin Material. *Trans. Chin. Electrotech. Soc.* **2016**, *31*, 24–30.
2. Zhang, X.; Wu, Y.; Wen, H.; Chen, X.; Xiao, S. Study on Microcapsules for Self-healing System of Insulating Epoxy Resins. *Proc. CSEE* **2018**, *38*, 2808–2814.
3. Wang, Y.; Liu, H.; Li, W.; Ding, R. An Investigation of Surface Tracking Characteristics and Factors Influencing Epoxy Resin Pouring Insulation for Dry-type Reactors. *Macromol. Res.* **2018**, *27*, 310–320. [CrossRef]
4. Liang, M.; Wong, K. Improving the long-term performance of composite insulators use nanocomposite: A review. *Energy Procedia* **2017**, *110*, 168–173. [CrossRef]
5. Wang, Y.; Liu, Y.; Wang, S.; Xu, H. The Effect of Electrothermal Aging on the Properties of Epoxy Resin in Dry-Type Transformer. *Trans. Chin. Electrotech. Soc.* **2018**, *33*, 250–260.
6. Arronche, L.; Saponara, V.; Yesil, S. Impact damage sensing of multiscale composites through epoxy matrix containing carbon nanotubes. *J. Appl. Polym. Sci.* **2013**, *128*, 2797–2806. [CrossRef]
7. Saurín, N.; Sanes, J.; Bermúdez, M. Self-Healing of Abrasion Damage in Epoxy Resin–Ionic Liquid Nanocomposites. *Tribol. Lett.* **2015**, *58*, 4. [CrossRef]
8. Hu, P.; Li, C.; Chen, D. Cause Analysis and Countermeasure Study of Cracking Accident of Cable GIS Terminal Epoxy Casing. *Electr. Power Eng. Technol.* **2017**, *36*, 102–105.
9. Wang, Y.; Liu, Y.; Xiao, K. The effect of hygrothermal aging on the properties of epoxy resin. *J. Electr. Eng. Technol.* **2018**, *13*, 892–901.
10. Chen, L.; Tsao, T.; Lin, Y. New diagnosis approach to epoxy resin transformer partial discharge using acoustic technology. *IEEE Trans. Power Deliv.* **2005**, *20*, 2501–2508. [CrossRef]
11. Huang, M.; Zhou, K.; Yang, D.; Yang, M. Effect of On-Line Rejuvenation on Water Tree Propagation in XLPE Cables. *Trans. Chin. Electrotech. Soc.* **2016**, *31*, 176–182.
12. Brown, E.N.; White, S.R.; Sottos, N.R. Retardation and repair of fatigue cracks in a microcapsule toughened epoxy composite—Part II: In situ self-healing. *Compos. Sci. Technol.* **2005**, *65*, 2474–2480. [CrossRef]
13. Yang, M.; Zhou, K.; Wu, K.; Tao, W.; Yang, D. A New Rejuvenation Technology Based on Formation of Nano-SiO_2 Composite Fillers for Water Tree Aged XLPE Cables. *Trans. Chin. Electrotech. Soc.* **2015**, *30*, 481–487.
14. Wei, G.; Tang, J.; Zhang, X.; Lin, J. Gray intensity image feature extraction of partial discharge in high-voltage cross-linked polyethylene power cable joint. *Int. Trans. Electr. Energy Syst.* **2014**, *24*, 215–226. [CrossRef]
15. Xu, T.; Dong, X.; Li, R.; Liu, G.; Xia, H.; Xia, Z. Ultrasonic Phased Array Detection of Composite Insulator Internal Defects. *Electr. Power Eng. Technol.* **2018**, *37*, 75–79.
16. White, S.R.; Sottos, N.R.; Geubelle, P.H.; Moore, J.S.; Kessler, M.R.; Sriram, S.R. Autonomic healing of polymer composites. *Nature* **2001**, *409*, 794–797. [CrossRef] [PubMed]
17. Bekas, D.G.; Tsirka, K.; Baltzis, D.; Paipetis, A.S. Self-healing materials: A review of advances in materials, evaluation, characterization and monitoring techniques. *Compos. Part. B Eng.* **2016**, *87*, 92–119. [CrossRef]
18. Kang, J.; Tok, J.B.-H.; Bao, Z. Self-Healing Soft Electronics. *Nat. Electron.* **2019**, *2*, 144–150. [CrossRef]
19. Li, W.; Dong, B.; Yang, Z.; Xu, J.; Chen, Q.; Li, H.; Xing, F.; Jiang, Z. Recent Advances in Intrinsic Self-Healing Cementitious Materials. *Adv. Mater.* **2018**, *30*, 1705679. [CrossRef]
20. Murphy, E.B.; Wudl, F. The world of smart healable materials. *Prog. Polym. Sci.* **2010**, *35*, 223–251. [CrossRef]
21. An, S.; Lee, M.W.; Yarin, A.L.; Yoon, S.S. A Review on Corrosion-Protective Extrinsic Self-Healing: Comparison of Microcapsule-Based Systems and Those Based on Core-Shell Vascular Networks. *Chem. Eng. J.* **2018**, *344*, 206–220. [CrossRef]
22. Zhang, X.; Ji, H.; Qiao, Z. Residual stress in self-healing microcapsule-loaded epoxy. *Mater. Lett.* **2014**, *137*, 9–12. [CrossRef]

23. Lesaint, C.; Risinggard, V.; Hølto, J.; Saeternes, H.H.; Hestad, Ø.; Hvidsten, S. Self-healing high voltage electrical insulation materials. In *2014 IEEE Electrical Insulation Conference (EIC)*; IEEE: Piscataway, NJ, USA, 2014; pp. 241–244.
24. Yang, Y.; He, J.; Li, Q.; Gao, L.; Hu, J.; Zeng, R.; Qin, J.; Wang, S.; Wang, Q. Self-healing of electrical damage in polymers using superparamagnetic nanoparticles. *Nat. Nanotechnol.* **2019**, *14*, 151. [CrossRef] [PubMed]
25. Wang, Y.; Li, Y.; Zhang, Z.; Zhang, Y. Effect of Doping Microcapsules on Typical Electrical Performances of Self-Healing Polyethylene Insulating Composite. *Appl. Sci.* **2019**, *9*, 3039. [CrossRef]
26. Schaubroeck, D.; Brughmans, S.; Vercaemst, C.; Scheaubroeck, J.; Verpoort, F. Qualitative FT-Raman investigation of the ring opening metathesis polymerization of dicyclopentadiene. *J. Mol. Catal. A Chem.* **2006**, *254*, 180–185. [CrossRef]
27. Álvaro, G. Self-healing of open cracks in asphalt mastic. *Fuel* **2011**, *93*, 264–272.
28. Wang, Y.; Wang, S.; Lu, G.; Huang, Y.; Yi, L. Influence of Nano-Al N Modification on the Insulation Properties of Epoxy Resin of Dry-Type. *Trans. Chin. Electrotech. Soc.* **2017**, *32*, 174–180.
29. Wu, Z.; Wang, C.; Zhang, M.; Pei, X.; Jiang, P. Interface of Epoxy Resin Composites, and Its Influence on Electrical Performance. *Trans. Chin. Electrotech. Soc.* **2018**, *33*, 241–249.
30. Chen, Y.; Meng, G.; Dong, C. Review on the Breakdown Characteristics and Discharge Behaviors at the Micro & Nano Scale. *Trans. Chin. Electrotech. Soc.* **2017**, *32*, 13–23.
31. Rohde, B.J.; Le, K.M.; Krishnamoorti, R.; Robertson, M.L. Thermoset Blends of an Epoxy Resin and Polydicyclopentadiene. *Macromolecules* **2016**, *49*, 8960–8970. [CrossRef]

Article

Is the Viscoelastic Sheet for Slamming Impact Ready to Be Used on Glass Fiber Reinforced Plastic Planning Hull?

Patrick Townsend [1,*], Juan Carlos Suárez Bermejo [2], Paz Pinilla [2] and Nadia Muñoz [1]

1 ESPOL Polytechnic University, Escuela Superior Politécnica del Litoral, Facultad de Ingeniería Marítima y Ciencias del Mar (FIMCM), Guayaquil 09-01-5863, Ecuador; nmunoz@espol.edu.ec

2 School of Naval Arquitecture and Marine Engineering (ETSIN), Universidad Politécnica de Madrid (UPM), 28040 Madrid, Spain; juancarlos.suarez@upm.es (J.C.S.B.); paz.pinilla@upm.es (P.P.)

* Correspondence: ptownsen@espol.edu.ec; Tel.: +593-991527396

Received: 14 July 2020; Accepted: 24 August 2020; Published: 19 September 2020

Abstract: Planing hull vessel built with polymer matrix laminates and fiberglass reinforcements (GFRP) suffer structural damage due to the phenomenon of slamming during navigation, due to the impact of the boat hull on the free surface of the water at high speed. A modification in the manufacture of the laminates for these fast boats is proposed, consisting of the insertion of an additional layer of a hybrid material, formed by elastomer encapsulated in an ABS polymer cell. Using GFRP specimens made from pre-impregnated material and reproducing the characteristic impacts of slamming, it is possible to compare the modified material with the introduction of the viscoelastic layers with the response under the same conditions as the unmodified laminates. Additionally, the panels have been tested using impacts due to weight drop at different energies, which allow determining the material damage threshold as a function of the energy absorbed, and to establish a comparison with the GFRP panels modified by observation in fluorescent light. It is verified that the proposal to reduce the effect of these impacts on the generation of damage to the material and its progression throughout the service life of the vessel is effective.

Keywords: slamming; damage; viscoelastic layer; prepreg; OoA

1. Introduction

Slamming is an important event during the navigation of the ship, and it appears as a sudden force that vertically impacts the ship in the bow and generates energy from the impact between the hull and the water surface. This force translates into pressure that acts on a very small surface and is so unpredictable that it still requires investigation [1]. This impact and its damage to the ship is so important, that the sailors are very cautious, they reduce the speed so as not to suffer additional damage during the voyages. The complexity of the phenomenon is due to the fact that the fluid enters the bottom of the ship due to the angular difference between the body surface, expanding at high speed. This generates very high pressures that are very important issues in the design of the ships. The answers about this phenomenon and its influence on the structure of the vessels have not been fully resolved and it is more complicated when they are GFRP planning vessels [2,3].

Tests have been carried out with complete models that try to simulate the real scale of the effect of slamming on the boat [4]. Experiments with complete ship models look for the global answer, it is quite expensive and to this must be added computational models and long-term simulations that try to explain the damage caused by the pressure whip in the ship's material. This is directly related to their premature aging. For this reason, mathematical models and reproductions of the fatigue event on the ship's hull are an important option for analyzing the response of the impacted structure [5].

The material used in this work is that of ships constructed of GFRP based on pre-impregnation cured in the oven. The phenomenon of slamming for these materials has the particularity that the impact of the sea is converted into energy that dissipates, producing different levels of damage, being one of the most important parameters in the design of the ship [6]. The purpose of the investigation is to reduce the damage produced on composite material by slamming impacts, so that it does not expand within the laminate. And prevent it from skipping between the layers, causing intralaminar and interlaminar damages that change the flexural stiffness.

The use of viscoelastic materials has been an option to try to absorb noise and impact on structural surfaces. Its use in GFRP vessels has been studied in some ways in the laboratory, to observe its benefits in energy dissipation [7]. Protection against damage energy was successfully demonstrated as it takes advantage of the fact that composite materials have high stiffness ratios and moderate level of damping. Combining high levels of energy dissipation with minimal structural stiffness achieves good results [8].

Elastomers with their Poisson constant property close to 0.5 allow them great elongations and energy absorption when restrained, therefore they are the key to the design of the viscoelastic sheet. By encapsulating the elastomer within a rigid polymer, it allows to play with the weight of the sheet, its adhesion capacity and its thickness. Considering that it will be placed in the hull of the boats, the benefits of not increasing the weight of the ship is interesting. A hexagonal design is proposed, which allows creating a set of cells that are grouped together with each other when manufactured and can form an easy to apply set.

Taking advantage of the properties of viscoelastic, Hooke's Law in three dimensions shows that viscoelastic must have only one exposed and free surface for compression, and its other two restricted directions. In such a way that the proposed viscoelastic layer has exposed the elastomeric material to be compressed inside a capsule. As a rigid plastic, ABS was chosen in previous investigations, which is very common in the market. As a linear elastomer, the TPU type is also widely used today. The proposed analysis of comparing cyclical slamming pressures with vertical impact is directly related to the requested compression.

The viscoelastic properties of these materials, confined in a less deformable material, contribute in a very significant way to improve the performance of the material in service and dissipate a greater percentage of the energy received on impact, minimizing the damage generated in the material due to the slamming pressure peaks generated during navigation. For this reason, the energy produced in each of the impacts is studied in the laboratory from the perspective of the vertical impact due to drop weight, introducing an energy accumulated in the reproduction of slamming. This added to the microstructural observation, shows significantly the behavior of the damage evolution [9,10]

Among the questions pending in previous studies, was the problem of the adhesion of the viscoelastic sheet on the matrix. Through cohesive theory model studies, this concern has been answered for designers, who must consider within their structural calculations that the stress and strain limits do not exceed the adhesion and spread propagation thresholds.

The test performed is a modified version of the ISO 11343, a standardized Wedge Peel test method that allows to measure the dynamic resistance of structural adhesives to cleavage at different strain speeds. In the application that is envisaged for this viscoelastic layer in GFRP, designed for protection of the hull of high-speed crafts against slamming impact events, it is important to measure the adhesion strength between the protective layer and the laminate. The test used for this purpose consist of a wedge with an acute shape being driven at a defined velocity into the adhesive bond between both substrates. The wedge induces a bending causing the bond to fracture and adherends to peel apart. The cleavage force and the dynamic resistance to cleavage are consequently calculated from the test force-displacement curve. Because the method is basically grounded in the mechanics of bending beams, it can be analyzed quantitatively using standard fracture mechanics analysis.

The absorbed energy is considered as a representative parameter of the behavior of composite materials when subjected to impact loads, and to study the energy behavior during impact, accelerometers with a computer data acquisition system have been used to quantify the energy returned. By observing impact forces and displacement, initial deformation and restitution are analyzed [11,12].

With the tests presented and the results are expected to answer the question designers if the viscoelastic sheet is ready for use in the construction of new boats GFRP. Energy parameters are also established to define how much the useful life of the planing hull vessel will improve, at the discretion of the manufacturers and users.

The present work is an extension of the previous investigations of the authors of this paper, referring to the use of the viscoelastic layer to improve the performance in the GFRP. The tests carried out in previous studies are presented in this article with complements that try to solve the new questions formulated by the designers: is it feasible to install the viscoelastic sheets in the GFRP planing vessel? Are they protected from the cyclical impact of slamming according to the previously resolved benefits and shortcomings? [13,14] For this reason, the cyclical slamming tests are complemented with the vertical impact due to weight drop to present its benefits from the perspective that the designer wishes to observe. The research aims to show another type of mechanical response of the viscoelastic layer under the perspective of vertical impact, and that the builders have different options for its use. The sheets are inexpensive and easy to cure in the construction of planing hulls.

The work has been organized according to the flow of activities shown in Figure 1, in order to present the experimental results from the definition of the experimental method, the slamming pressure levels to be used in vertical impacts, and with this information present discussion in which the comparison of results shown.

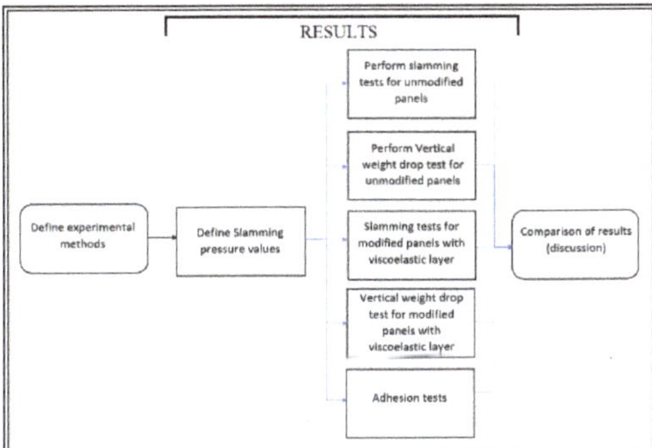

Figure 1. Organization of the research presented.

2. Materials and Methods

The reproductions of slamming impact were made with the equipment presented in Figure 2. It is composed of a variator-electric motor assembly coupled to a cam with a shaft supported by two flexible supports to dissipate the reaction energy of the cam when pressing the panel and protect the drive from possible excessive lateral loads. The cam is keyed while remaining fixed to the shaft and rotates between 200 to 320 RPM. The panel was installed at its base with gauge screws that allowed exerting pressures between 0 to 1200 kN/mm^2 and the panel is adjusted crosswise to the cam. A revolution counter was installed, and the temperature control was carried out with a thermal imager, to avoid that the temperature does not reach the glass transition temperature of the material due to friction.

Two types of panels were made of GFRP, without viscoelastic layer which was called "unmodified" and with viscoelastic layer which was called "modified". Some were made from Gurit WE-91 triaxial (0°/45°/90°) pre-impregnated OoA (Out-of-Autoclave curing), cooled to −18 °C, vacuum cured. The panels were manufactured using 3 fabrics of 1 mm each, cut to 270 × 270 mm corresponding to the frame dimension of the test equipment. In the panels with the viscoelastic sheet, it was included after the first triaxial sheet. All the panels included a strain gauge at 40 mm from the impact zone.

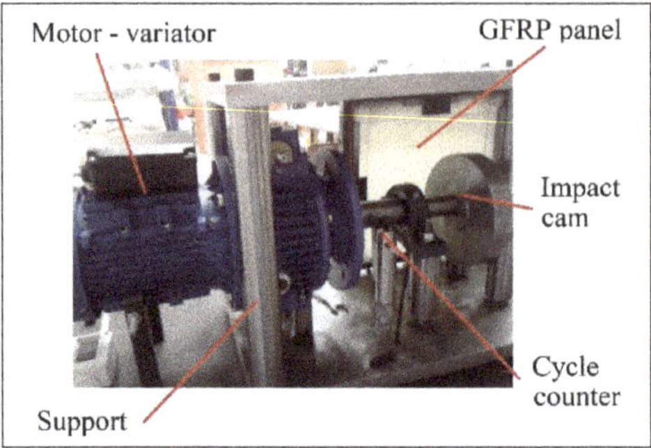

Figure 2. Reproduction equipment for slamming impacts.

For the manufacture of the viscoelastic sheets, a printer was used with two independent extruders so that there is no contamination between the materials. They were printed on a thermal surface. For the manufacture of the outer capsule, 3 mm acrylonitrile butadiene styrene or ABS was used. A linear thermoplastic polyurethane or TPU was used for the inner elastomer. Figure 3 shows a generic viscoelastic layer. Depending on the manufacturing date for the experiments, the color type of the ABS varies, so there are black and blue.

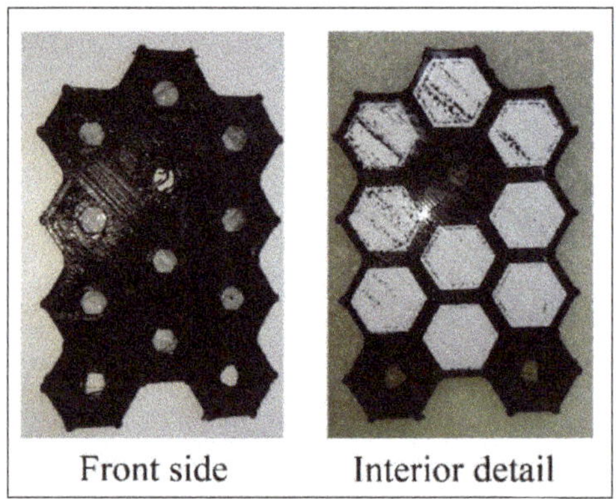

Figure 3. Generic viscoelastic layer.

The curing process of the panels was with a vacuum pressure bag and in an oven at a temperature of 120 degrees Celsius for 90 min. This process was applied to unmodified panels as well as modified panels.

A 12-m-long vessel constructed from GFRP was selected, and the maximum allowable slamming pressure values were calculated. This pressure occurs from the midsection towards the bow at the bottom of the ship. The purpose was to perform the slamning reproductions with lower values, in such a way that the failure of the material is due to propagation of the damage due to the cyclic load. The ABS classification rules for high-speed boats, Chapter 3, Section 2.2 [13] were used for the calculation. The ship's bottom pressures were calculated for the fully loaded condition, operating at maximum speed according to the selected geographic navigation area: The Galapagos Islands.

Three-point bending tests were performed to experimentally determine the pressure threshold from which damage begins to occur on the unmodified material. Cured specimens were cut from a panel into strips 15 mm wide by 250 mm long. They were flexed beyond their elastic limit until the tension machine registers non-linearity, indicating the appearance of the first micro-cracks equivalent to the damage threshold as seen in Figure 4.

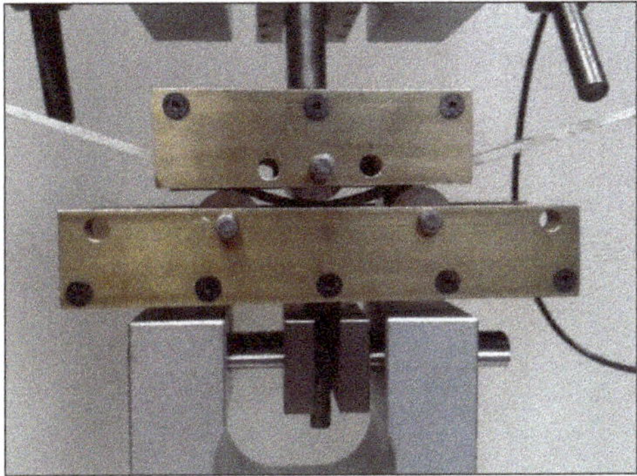

Figure 4. Three-point Bending test.

A machine for vertical weight drop testing was built. The impactor carriage falls by gravity on two chrome rails to reduce the effect of friction. The panels to impact were installed at its base, with all its edges embedded. It has an anti-bounce system with a laser reader to control the number of impacts. It has an acceleration sensor or gravitometer installed that sends the information to a data acquisition system and tabulates the acceleration G-force versus the time of impact. The equipment has a frame-shaped structure to which all parts are secured. The impactor was made with electromagnets for holding and launching. The guide bolts allow more weight to be added to it. At the bottom of the impactor it has the impact tip with a magnetized sphere. The number of impacts is controlled with a laser reader. Figure 5 shows the detail of the impact equipment and its parts.

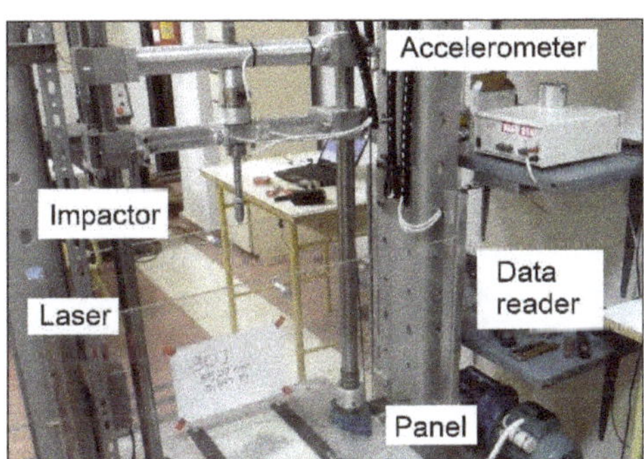

Figure 5. Weight drop impact equipment.

With the data on the forces of gravity read by the accelerometer, the following formulation is applied to calculate the energy absorbed and energy returned.

G_i: dimensionless number of gravities i given by the gravitometer.

For each Gi data reading the following parameters were calculated:

Δt_i: time interval (s)
M: impactor mass (kg)
f_i: impact force (N)
x: impactor displacement (m)
k_i: kinetic energy of impact (J) received by the impactor

E_i: impact energy (J) delivered by the impactor
E_o: initial energy of the impactor (J)
$E_{a\,i}$: energy absorbed by the material (J).

The results of the impact tests were processed with the following formulation to obtain the values under the curves of impacted energy and absorbed energy, and to be able to estimate the energy that transforms into damage.

$$f(t) = 9.81{*}M{*}G_i \tag{1}$$

$$x_i = \frac{3}{2} * 9.81 * G_i * \Delta t_i^2 \tag{2}$$

$$k_i = \frac{1}{2} * 9.81^2 * G_i * M * \Delta t_i^2 \tag{3}$$

$$E_i = \frac{3}{2} * 9.81^2 * G_i * M * \Delta t_i^2 \tag{4}$$

$$E_{a\,i} = E_o - (|E_i| - |k_i|) \tag{5}$$

The impact tests were carried out in different energy ranges, varying the impactor weight and height.

To observe micro cracks within the laminate after impact and to make a comparison between unmodified and modified panels, characterization was performed with fluorescent penetrating inks to expose the sections to ultraviolet light with the procedure presented in Figure 6. The panels in the impact area were cut from 60 × 60 mm and drilled with a 0.5 mm drill bit, for the purpose of immersing them in a fluorescent penetrating liquid. Sectional cuts were made in the specimens, to observe their different intralaminar delaminations in interlaminar with fluorescent light. In the case of panels with viscoelastic sheet, this was withdrawn during exposure, because it is important to compare the viscoelastic layers protected.

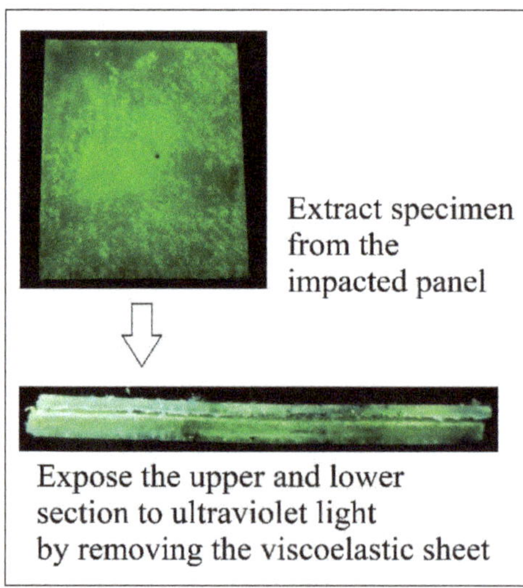

Figure 6. Test evaluation procedure for fluorescent penetrating inks.

In order to know the adhesion force that the viscoelastic sheet has once cured in the modified panels, shear tests adjusted to the Cohesive model theory were performed to estimate the damage initiation phase force, the maximum force at which detaches the sheet and the behavior of the propagation force of the detachment. For the shear tests, single-layer panels of Mat 200 were manufactured. This allows the blade to work only on the viscoelastic sheet and the matrix as seen in Figure 7, the universal testing machine was used with a very fine steel blade, which will make the cut parallel to the viscoelastic.

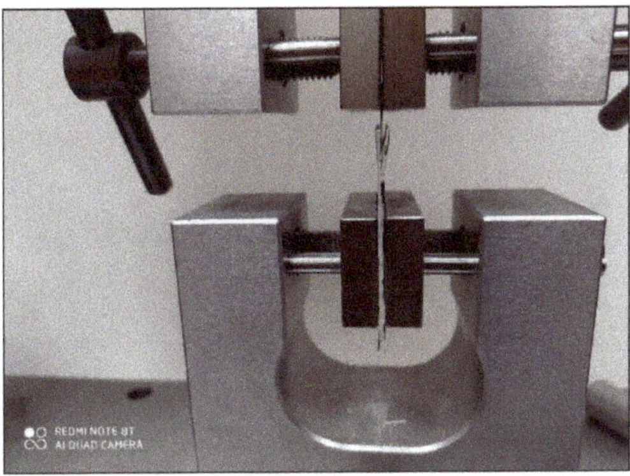

Figure 7. Shear test equipment.

3. Results

3.1. Slamming Pressure Values

For the calculation of slamming pressures according to the ABS classification rules, the typical GFRP vessel with length between 12 to 15 m and sailing at a nominal speed of 22 knots was considered. The site of operation is in the Galapagos Islands with an ocean of Beaufort scale 4, that is to say with waves of height of 1.5 to 1.8 m and frequencies of 14 to 19 s of bottom sea. The maximum design pressure at the bottom of the boat was between 1800 kN/m^2 to 1050 kN/m^2.

From the results of the three-point bending test to determine the damage threshold of the laminate, it was determined that over the 2300 μm/m of deformation, the curve was no longer linear and corresponded to the desired threshold.

Using the Finite Element Method (FEM), the force that caused deformations equivalent to those that cause damage to the material was estimated in the model. Figure 8 shows the symmetric model with the load at its center and a detail of the section in the impact zone of the same model. The results of the elastic strain have been plotted for a value in the maximum impact zone of 834 μm/m corresponding to a pressure of 385 kN/m^2.

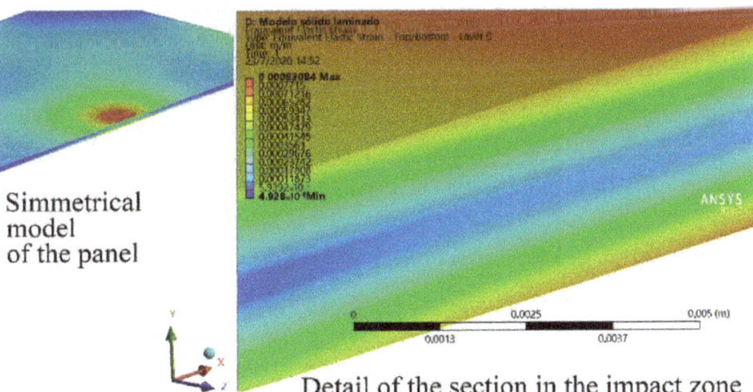

Figure 8. FEM model.

The pressure range in which the slamming tests could be carried out was determined to be 260 to 820 kN/m^2. The same procedure was applied to calculate the damage threshold pressure, which was obtained from 358 kN/m^2.

3.2. Slamming Tests for Unmodified Panels

Table 1 shows the tests carried out on the unmodified panels. The variables presented are the slamming pressure in kN/m^2 applied by the equipment on the contact surface, the number of cycles applied and the frequency in RPM of the impacts. The frequency variation depended on the motor-variator, so that the pressure on the cam does not produce loads that affect the system.

Table 1. Variables used in unmodified panels for slamming tests.

Panel #	Pressure (kN/m^2)	Cycles ($\times 10^5$)	Frequency (RPM)
A	260	2.10	210
B	400	1.50	220
C	630	1.81	220
D	810	0.21	310
E	830	0.21	310
F	420	0.27	310

In the 810 kN/m^2 and 830 kN/m^2 panels, the number of cycles of each group in these panels depended on the temperature of the cam which, due to friction, reached 70 °C. These panels had greater damage, in a low number of cycles of the order of 2×10^4.

Panel D reached 85% damage in the area of contact with the cam. It was observed that the first micro cracks appeared at 2×10^2 impacts and were already highly visible at 1×10^2 cycles. The first damage observed as slight white shadows were in the areas where the sides of the cam meet. They then lined up toward the center of the contact surface and show the evolution of damage as slamming strokes increase. Figure 9 shows the results of the test.

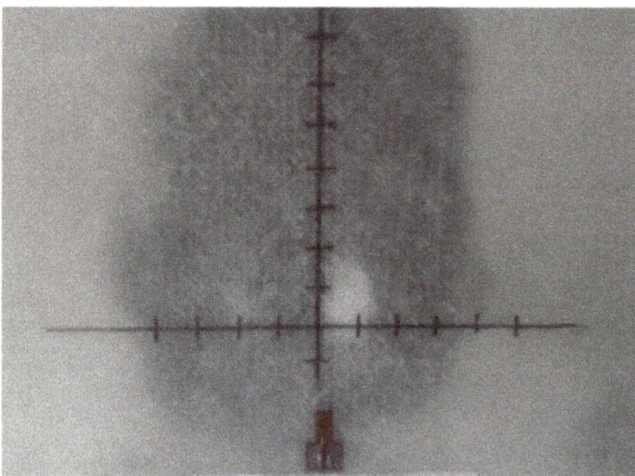

Figure 9. Evolution of unmodified panel D damage at 0.21×10^5 impact cycles with P = 810 kN/m^2.

Damage spread could be observed in real time at this pressure level and was first expanded by the penultimate laminated −45° layer on the strain gage side. At the end of the test it was found that breaks had already formed in the panel that passed from side to side in the laminate. Normalizing the percentage of damage between the maximum damage area value for each test in panels A–E. It is observed in Figure 10 that the evolution of the damage has a similar behavior for low energy impacts, and similar behavior for high energy impacts. This according to the pressure values presented in Table 1 values used in the slamming tests of unmodified panels.

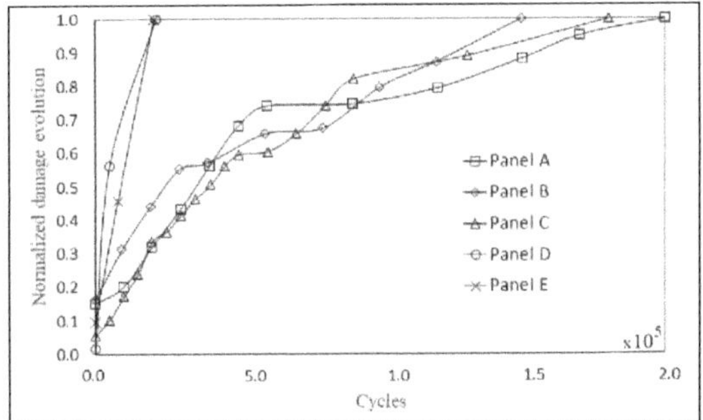

Figure 10. Normalized damage evolution on unmodified panels.

The change in flexural stiffness was evaluated with the results of the strain gage using panel F. In Figure 11 are the results of the for a deformation of 1×10^3 µm/m at the slamming pressure point in the 0.27×10^5 test cycles with a pressure of 420 kN/m^2. The data has a representative ascending line, which indicates that as the micro cracks increase, the panel loses rigidity. This is in accordance with the type of damage observed on the panel on the opposite side of the impact. These values were calculated from the deformations obtained with the strain gages.

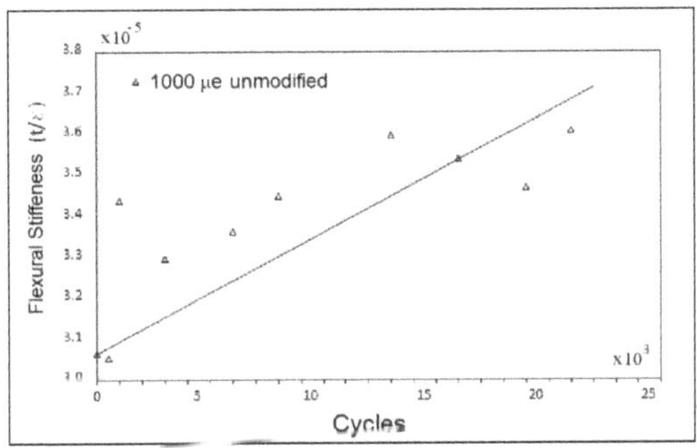

Figure 11. Evolution of flexural stiffness in panel unmodified F up to 0.27×10^5 impact cycles with P = 420 kN/m^2.

These results will be considered for the relationship between low energy slamming impacts and vertical weight drop tests. It is clearly identified in the results that some values presented in the graph correspond to high energy tests, and other tests correspond to low energy tests. Composite material has a defined energy absorption boundary.

3.3. Vertical Weight Drop Tests for Unmodified Panels

The impact tests were carried out in different energy ranges for a certain number of rebounds, and with the strain gage the maximum deformation of the panel was measured. These results are shown in Table 2.

Figure 12 shows the sequence of damage to the panels on the impact side, for the different energies tested with vertical weight drop. The sequence allows visualizing the increase in damage that occurs as the applied energy increases.

Table 2. Variables used in unmodified panels for vertical weight drop tests.

Impact (J)	Maximum Deformation (μm/m)	# of Rebounds
10	502	1
20	749	1
30	1130	1
40	1401	1
50	1710	1
60	1828	1

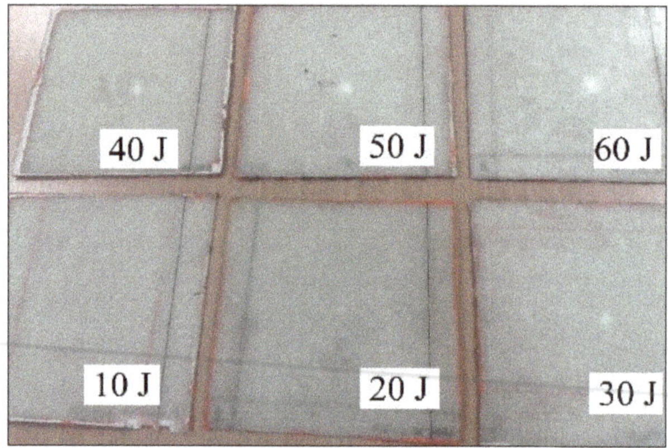

Figure 12. Unmodified panels impacted by vertical weight drop test from 10 J to 60 J of energy.

As the impact energy was increased, the microcracks were observed to be oriented in the direction of the fiber. The largest number of observed micro cracks evolved over the 45° layer. To a lesser extent in the 0° layer. Impacts less than 40 Joules can be considered low energy, while those greater than 40 Joules high energy, where the panels already show delamination and deep separation of layers. The phenomenon of slamming is a low energy process, for which the panels impacted at 30 J are the best to be considered for comparison with the panels tested in fatigue.

When characterizing the panels tested with penetrating inks, it was observed that they had indeed entered the delamination, being easily seen under ultraviolet light. Sections sequentially are exposed to fluorescent light as seen in Figure 13 for impact at 40 Joules respectively, in which interlaminar delamination can be easily observed.

Figure 13. Unmodified panel sections exposed to ultraviolet light impacted at 40 J.

As expected in a laminated composite material, the breaks in the matrix are vertical and connect with delamination between layers, producing a ladder shape. The breaks observed in the lower layers indicate that the impact energy was converted into damage and was jumping from one layer to another. It is also important to note that according to the orientation of the layers, the delamination moves from left to right starting from the vertical direction of impact.

The accelerometer records the acceleration force (G), at a frequency of 104 Hz. According to the suggested equations, they show that the acting force versus displacement has the profiles presented in Figures 14 and 15. The peaks of damage that occurs as the impactor breaks each of the layers. The impactor slows down each time a new delamination occurs. Records the acceleration variations when micro cracks occur in the matrix. In the case of the low energy impact of 30 Joules, the impact force reaches its maximum value by breaking layer 5. Thereafter the layers offer resistance to deformation. In the case of impact at 50 Joules of high energy, the impactor has enough force to continue breaking all the layers until 9 at which it has already bounced.

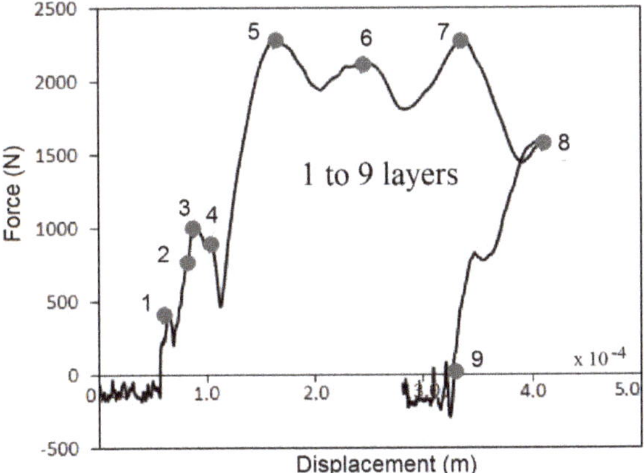

Figure 14. Comparison of the force vs. displacement diagram for an impact at 30 J in unmodified panel.

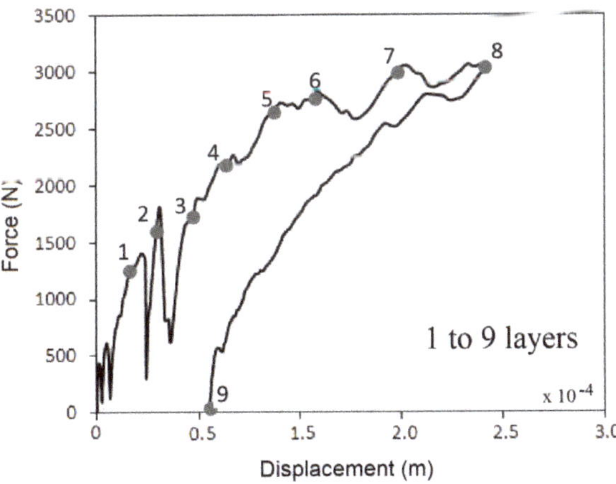

Figure 15. Comparison of the force vs. displacement diagram for an impact at 50 J in unmodified panel.

The acceleration curve also allows you to compare how fast laminate damage is and which layers absorb the most damage energy. In the case of the 30 Joule curve, between layers 4 and 5 there is a large increase in damage energy, indicating that this layer broke and delaminated. This is in accordance with what was observed in the sections of the material exposed to penetrating ink and ultraviolet light. In contrast to the panel at 50 Joules, the high energy impact imposes a lot of kinetic speed, and the first layers up to 6 breaks at the same time interval.

The assessment of the absorbed energy is observed in Figure 16 for impacts of 20, 30 and 40 Joules. The curves clearly show that the panel absorbs energy that is returned in the form of kinetic energy, and over a certain time interval its capacity to absorb energy decreases, indicating that there is more damage within the panel.

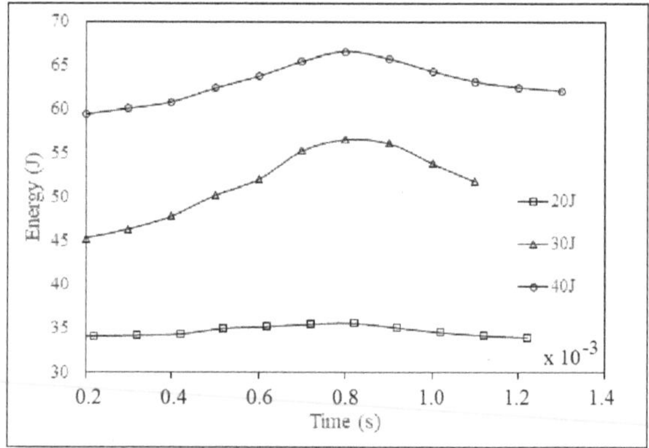

Figure 16. Energy absorbed by vertical energy impacts on unmodified panels.

The absorbed energy curve that has the best configuration is the 30 Joule curve, while the 60 Joule curve has an uncharacteristic trend, which is correctly related to the detachment and breakage of the matrix and the fiber that was observed in the test.

In the 20 Joule curve, its little variation is consistent with that observed in the panel after impact. The microcracks and directions of the delamination are very light. This indicates that little energy was transformed into damage and most was returned to the impactor.

3.4. *Slamming Tests for Modified Panels with Viscoelastic Layer*

The slamming tests carried out on viscoelastic modified panels are shown in Table 3. They correspond to the G panel test with high slamming energy whose microdeformation is close to the damage threshold. And the tests with panels H, K and L with cyclic impacts at low energy. The frequency variation depended on the motor-variator, so that the pressure on the cam does not produce loads that affect the system.

Table 3. Conditions of the slamming tests to modified panels with viscoelastic sheet.

Panel #	Pressure (kN/m^2)	Cycles (×10^5)	Frequency (RPM)
G	801	0.20	302
H	343	0.27	309
K	302	0.27	320
L	295	0.27	320

During panel G tests it was observed that over the 1×10^3 cycles, the first microcracks in the matrix were observed on the viscoelastic sheet. It was a high energy test and at the end of the test no delamination or cracks were observed on its surface. Panels K and L showed little damage to the matrix, and these were analyzed for adhesion with shear tests. Panel K is shown in Figure 17, which at 1×10^3 impact cycles already shows the appearance of microcracks, whose damage evolution was not significant.

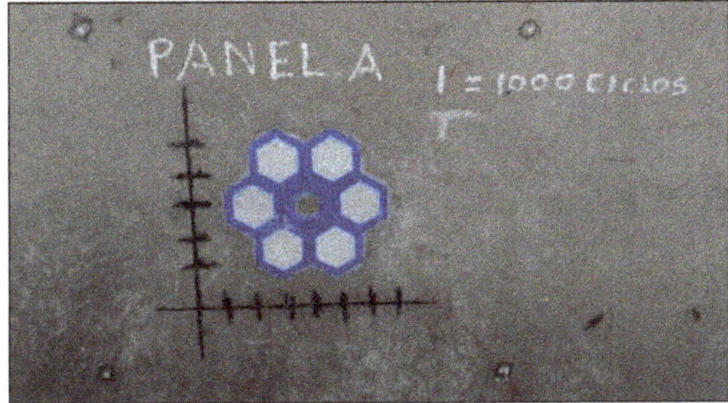

Figure 17. Modified panel K impacted at 1×10^3 cycles with 343 kN/m^2 slamming pressure.

The use of Mat fiberglass allowed the micro cracks that appeared inside the panel to be well visualized. During the test, the ambient temperature could be controlled so that there were no problems with the panel temperature. The control was made with the thermal imager and did not reach more than 50°. Also, the crystallinity of the resin, allowed to observe that there was no detachment of the viscoelastic during the test.

For the modified panel F, 2.7×10^5 impacts were made with a slamming pressure of 343 kN/m^2, evaluating its change in flexural stiffness. The change versus cycles is presented in Figure 18. It is observed that the representative line has a low magnitude slope and the data tends to change in a lower value. This is related to what was observed during the test, in which few microcracks and damage evolution after impact are observed. The protection of the viscoelastic layer is clearly observed because the damage is minimal on the face after impact. The orientation of fiber damage is not clearly defined. The results are considered good because they have an acceptable tendency, and in high energy levels there is also a concordance in flexural stiffness.

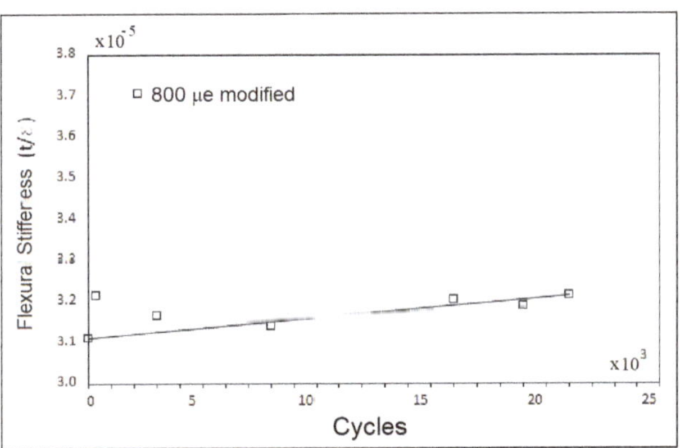

Figure 18. Evolution of flexural stiffness in modified panel F up to 0.27×10^5 impact cycles with P = 343 kN/m^2.

3.5. Vertical Weight Drop Tests for Modified Panels with Viscoelastic Layer

The tests for vertical drop in weight were carried out in different energy ranges, varying the impactor weight and height as seen in Table 4. This in order to observe results for low energy and high energy tests.

Table 4. Conditions of the vertical weight drop tests on panels modified with viscoelastic layer.

Impact (J)	Deformation Maximum (μm/m)	# of Rebounds
20	252	1
40	655	1
80	991	1
120	1040	3

The tests carried out on the panels with a viscoelastic layer with impact energies for weight drop of 20 and 40 Joules are shown in Figure 19. The impact of 20 Joules presents little damage on the surface with little failure of adhesion with the laminate matrix. In contrast, in the 40 Joules panel, the damage is much greater and there is a complete cradle separating the viscoelastic lamina with the matrix. The impact is punctual and there is a noticeable difference with the high energy slamming test. The vertical weight drop tests obtained similar results for the low energy case. Over 30 Joules the data obtained is applicable for a comparison.

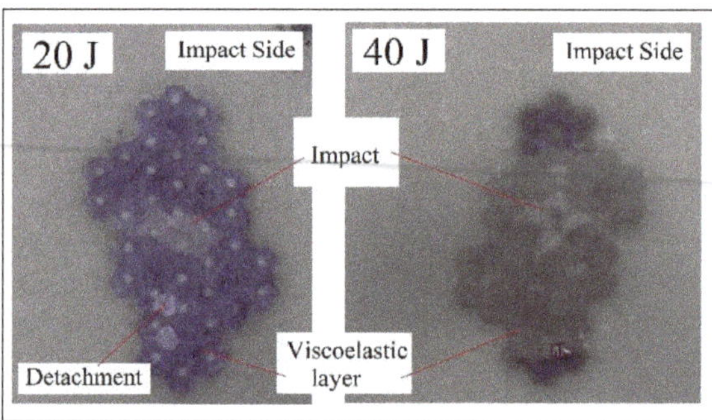

Figure 19. Impact tests for gravity weight drop of 30 and 40 J of energy to modified panels.

The modified panels were treated in the same way with fluorescent penetrating inks to assess the type of damage. During the cut, the viscoelastic sheet came off. This is seen in Figure 20. It should be noted that the viscoelastic layer came off due to the oil cut. The sections were 2 mm thick and despite the slow cutting speed, this affected their adherence. The cutting speed was very slow and oiled, to prevent fragments of composite material from sticking to the sections and showing erroneous results under fluorescent light.

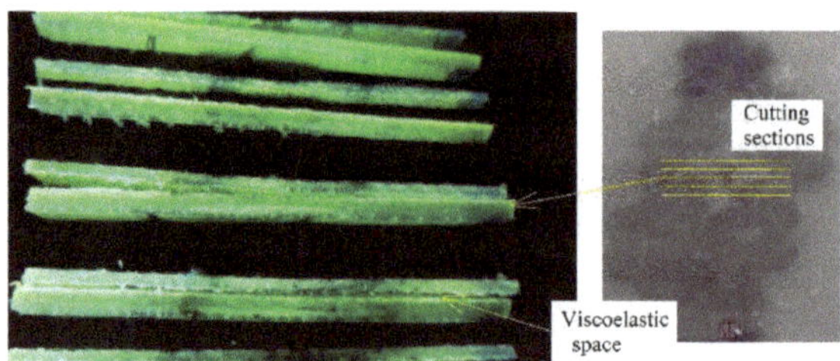

Figure 20. Characterization with penetrating ink to panels impacted at 40 J and 80 J.

It can be seen in the cross sections that the impacted laminate layers before the viscoelastic layer have damage to the matrix and fiber. The layers after this have little delamination. The matrix shows few vertical breaks and the ladder joints are not clearly visible. Delamination do not take a lateral direction due to the orientation of the layers but appear intermittently and hardly visible. All sections exposed to ultraviolet light show the same behavior. Vertical impact is similar to a fatigue event shown in slamming trials.

According to the results of the accelerometer and the use of the respective formulation, the behavior of the panel during the impact at 40 and 80 Joules is observed in Figures 21 and 22. It is observed that the impactor breaks layers 1, 2 and 3, and then experiences a decrease in force. This means that the impact force is damped by the viscoelastic layer. The impactor breaks the next layers, from the decreased force causing less damage. Depending on the impact force, the impactor does more or less damage to layer 9 on which it has bounced. Both results, the presence of the viscoelastic sheet is fully identified. At the 80 Joules high-energy impact, the layers broke at similar time intervals, until the interference of the viscoelastic layer. The acceleration curve is clearly marked as it affects the impactor and protects subsequent layers.

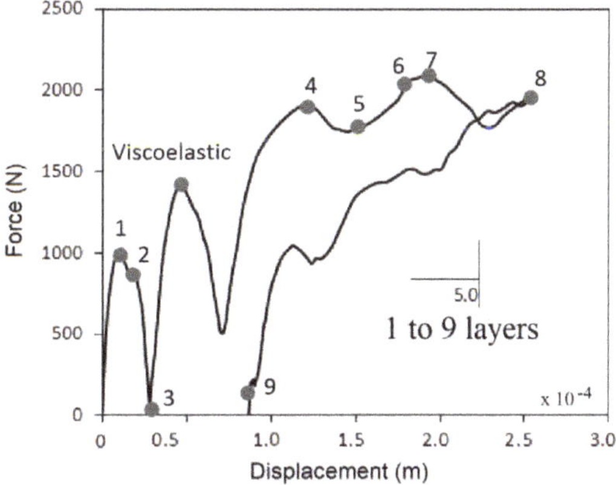

Figure 21. Comparison of the force vs. displacement diagram for an impact at 40 J in modified panel.

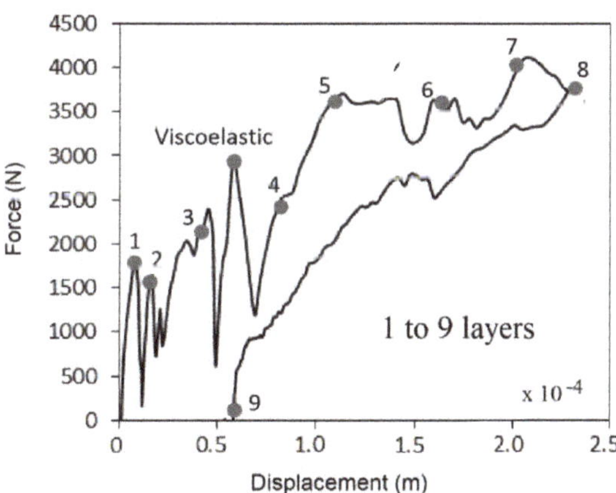

Figure 22. Comparison of the force vs. displacement diagram for an impact at 80 J in modified panel.

The energy absorbed by the modified panels for impacts of 30, 40, 80 and 120 Joules is shown in Figure 23. It is observed that the panels maintain their capacity to absorb impact energy, which indicates that the damage does not increase proportionally as which increases the impact energy. This is indicative of less damage to the matrix, which accumulates the kinetic energy imposed by the impactor. It is observed that at 120 Joules, the absorbed energy curve does not show the typical tendency of composite materials to high energy impacts. It does not have the ability to return all damage, but it does return damage energy to the impactor. This is identified in the negative slope in the intervals that the impactor reaches layer number 9. In such a way that the curve clearly shows that the material receives little damage for high energy vertical impacts. This result is unrelated to slamming energy impacts.

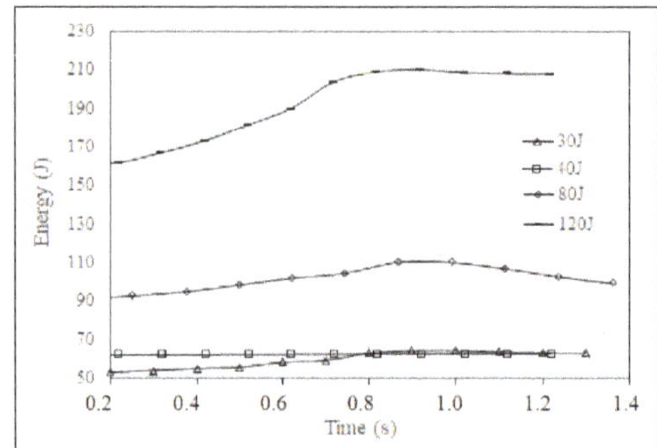

Figure 23. Energy absorbed by impact of weight drop tests on modified panels.

3.6. Adhesion Tests

Adhesion tests gave the expected results according to the "Cohesive Theory Model". Panels K and L were shear tested after being subjected to 0.27×10^5 slamming cycles. Figure 24 shows specimen K with the detachment of the viscoelastic. The tests were carried out on two sides in order to have more results to know the behavior of the adhesion of the sheet to the matrix.

Figure 24. Panel K tested with the shear test.

Special care was taken when the specimen was placed in the presses, so that it is in line with the blade. The speed of the machine was calibrated so that the test is carried out at the minimum capacity. In such a way that the results are representative of the different acting forces in the cohesive zone

The resulting forces that were obtained in the tests are shown in Figure 25 for panels K and L. The damage initiation phase is clearly identified in each of the tests, and there is little difference in the value of maximum force. That is required for the viscoelastic sheet to peel off. Once detached, the team was able to test the propagation of the detachment of the layer. The maximum force value corresponds to the shear force of the laminate so that the adhesion fails.

The results of the test were bin visualized in the damage of the test piece, because the use of the mat cloth allowed it to be crystalline. The blade when entering between the viscoelastic sheet and the matrix, could be followed to ensure that the test is well performed. The final data of the trial should correspond to the threshold of viscoelastic adhesion damage with an epoxy matrix.

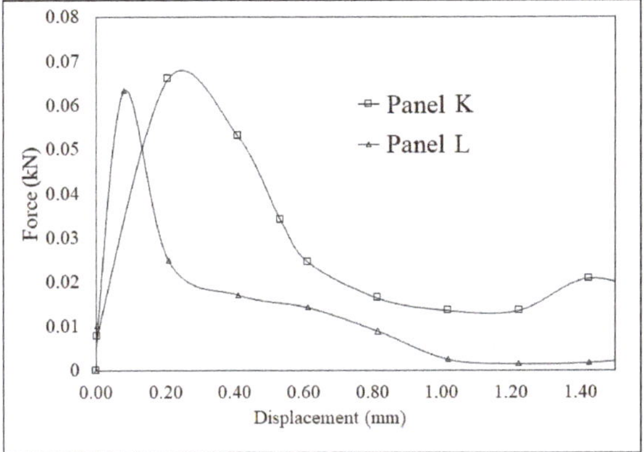

Figure 25. Adhesion force obtained with the shear test.

4. Discussion

If the damage caused by the impact of slamming that occurs in the hulls of the GFRP vessels could be observed in the tests. These do evaluate the evolution of the energy that is being converted into damage inside the laminate.

The insertion of a viscoelastic layer effectively mitigates the damage caused by the slamming phenomenon by reproducing the impacts under controlled laboratory conditions. The protection from the damage presented allows to absorb the dissipation of destructive energy, protecting the structure of the ship's hull and increasing its useful life. This brings a new perspective to the design of the ships and their scantlings, since the viscoelastic modification changes the way in which the stress concentrations are distributed in the hull of the boats. Comparing with the flexural stiffness values obtained along the scale of applied cycles, we observe in Figure 26 that over the 5×10^3, the unmodified and modified panels begin to distance in magnitude. The unmodified panel quickly gains flexural stiffness while the modified panel remains and does not change significantly.

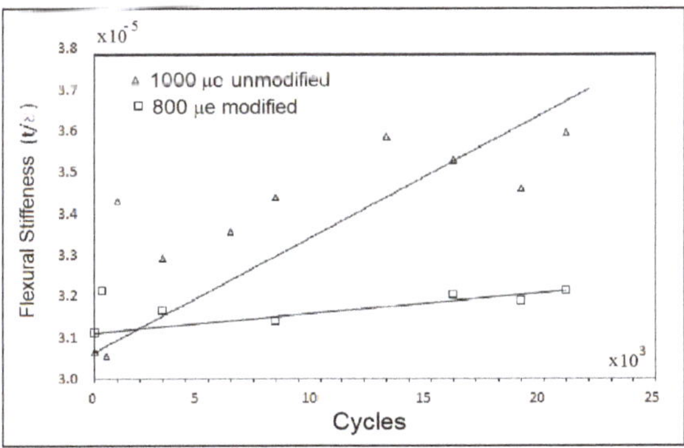

Figure 26. Comparison of the change in flexural stiffness during the slamming test with 0.27×10^5 cycles.

The panels impacted in pressure levels with values above and below the damage threshold calculated with the three-point flex test, marked an additional parameter to relate it to the maximum slamming pressure recommended by the ABS rules for the type of vessel selected.

Visual observation of the damaged surface within the contact area with the cam served to calculate the percentage of damage. This is related to the stress values that exceeded the damage threshold, transformed into microcracks.

The characterization by fluorescent penetrating inks, allowed us to physically observe how the damage does not propagate in the same way under the layers of the viscoelastic sheet, mixing this damage between the impact shock and the normal tension due to the flexion of the panel on the tensile side. Figure 27 shows two cut sections in the impact area already characterized by an unmodified and a modified panel, both impacted at 40 Joules. In the panel without modifying the interlaminar and intralaminar damages, they are linked, producing internal breaks in the laminate between layers 5 and 9 with important separations to the matrix. On the other hand, in the modified panel, the viscoelastic sheet protects the section of the laminate between layers 6 and 9, which correspond to those subsequent to the sheet. This means that it has returned the energy of damage and therefore little damage is observed on the underside, corresponding to microcracks and no delaminations.

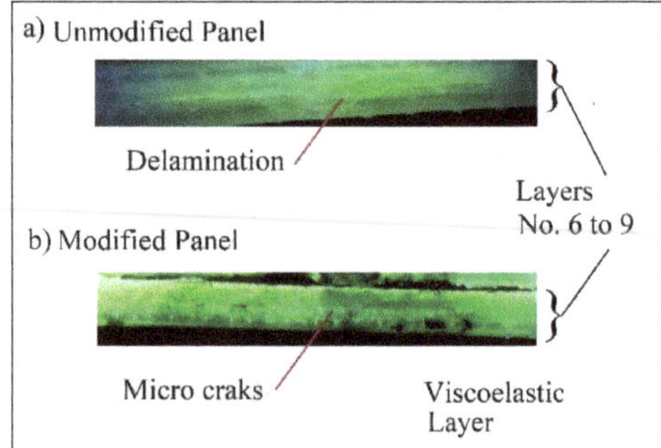

Figure 27. Comparison of the impact section for weight drop at 40 J characterized with fluorescent penetrating inks for unmodified panels (**a**) and modified panels (**b**).

Figure 28 shows the same comparison, but for a test of 80 Joules. Damage to the unmodified panel is very severe and unusable. On the other hand, the damage for the modified panel is less. If there is delamination of layers 6 to 9 that correspond to the layers after the viscoelastic layer, but not in all the layers. Viscoelastic layer protected these layers from severe impact damage.

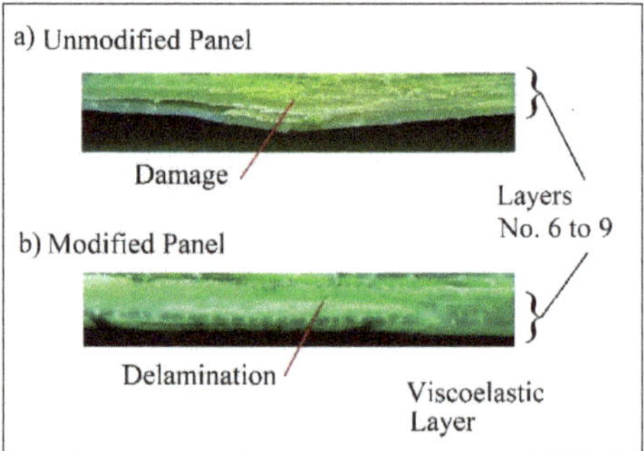

Figure 28. Comparison of the impact section for weight drop at 80 J characterized with fluorescent penetrating inks for unmodified panels (**a**) and modified panels (**b**).

According to these results, when comparing the absorbed energy calculations shown in Figure 29, there is a notable difference in the amount of energy for the same test at 30 Joules of impact. The modified panel absorbs little energy and returns it in a better way demonstrating protection.

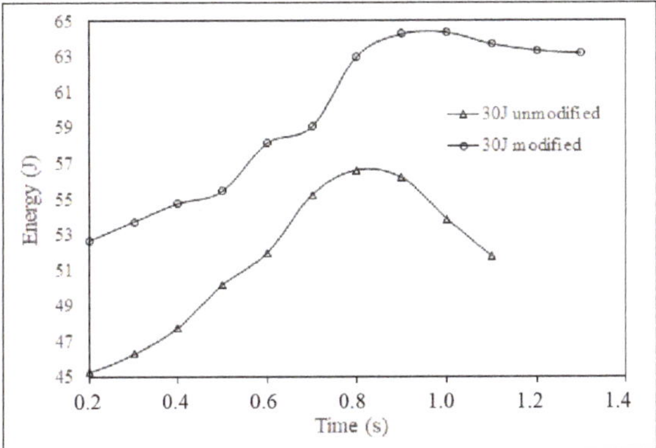

Figure 29. Comparison of energy absorbed by impacts of weight drop test.

5. Conclusions

The proposed methodology is a form of reproduction to assess damage during navigation, according to the slamming pressures that occur at the bottom of the hull of a boat. The pressures exerted on the GFRP for this case, is related to the percentage of damage produced per cycle with a tendency to increase pressure intensity. This is because the microcracks that align quickly with the fibers with the highest tension, are oriented to reach the fracture due to the decrease in stiffness and increased brittleness of the material. Penetrating ink tests confirm these results. The equipment is suitable for conducting low energy slamming impact tests in which the microcracks are aligned interlaminar for medium energy level impacts in which interlaminar damage also produces intralaminar damage. The impact energy was definitely absorbed by the viscoelastic layer. The viscoelastically modified GFRP protects itself from cyclical impact shocks from slamming, and the measure of maintaining the flexural stiffness is the ability to extend the life of the hull of ships made from this material. The adhesion values of the sheet allow designers to define their laminates so that the stresses of the material do not exceed the release values of the viscoelastic layer. The correct location of the viscoelastic sheets in the areas of greatest concentration of stresses due to the hit of the ship with the sea, is a starting point in the modification of shipbuilding to produce new types of more resistant ships. The viscoelastic sheet after this extensive investigation is ready to be inserted in the new constructions of planing hull vessels.

Author Contributions: In the research, P.T. carried out the execution of experiments, development of the formulations and narration of the manuscript. J.C.S.B. and P.P. collaborated with the conception and manufacture of the laboratory machines used in the research and the conceptual design of the experiments. N.M. with the conception of the GFRP application and revision of the manuscript. All authors have read and agreed to the published version of the manuscript.

Funding: This research received no external funding.

Conflicts of Interest: The authors declare no conflict of interest.

References

1. Kabsenberg, B. Slamming or ships: Where are we Now? *Philos. Trans. R. Soc.* **2011**, *369*, 2892–2897. [CrossRef] [PubMed]
2. Kapsenberg, G.; Veer, A. Whipping loads due to aft body slamming. In Proceedings of the 24th Symposium on Naval Hydrodynamics, Fukuoka, Japan, 8–13 July 2002.
3. Qin, Z.; Batra, R. Local slamming impact of sandwich composite hulls. *Int. J. Solids Struct.* **2009**, *46*, 2011–2035. [CrossRef]
4. Lavroff, J.; Davis, M.; Holloway, D. The whipping vibratory response of a hydroelastic segmented catamaran model. In Proceedings of the 9th International Conference on Fast Sea Transportation (FAST), Shanghai, China, 23–27 September 2007.

5. Lake, S.; Eagle, M. Slamming of Composite Yacht Hull Panels. In Proceedings of the 18th Chesapeake Sailing Yacht Symposium, Annapolis, MD, USA, 2–3 March 2007.
6. Centea, T.; Grunenfeldr, L. *A Review of Out-of-Autoclave Prepregs—Material Properties, Process Phenomena, and Manufacturing Considerations*; Composite Materials; Elsevier: Amsterdam, The Netherlands, 2015; Volume 70, pp. 132–158.
7. Kumar, S.; Ramesh, T. A continuum damage model for linear viscoelastic composite materials. *Mech. Mater.* **2003**, *35*, 463–480. [CrossRef]
8. Blake, J.; Shenoi, R. Strength modelling in stiffened FRP structures with viscoelastic inserts for ocean structures. *Ocean Eng.* **2002**, *29*, 849–869. [CrossRef]
9. Baucoma, J.; Zikryb, M. Low-velocity impact damage progression in woven E-glass composite systems Composites Part A. *Appl. Sci. Manuf.* **2005**, *36*, 658–664. [CrossRef]
10. Belingardini, G.; Bador, V. Low velocity impact tests of laminate glass-fiber-epoxy matrix composite material plates. *Int. J. Impact Eng.* **2002**, *27*, 213–229. [CrossRef]
11. Reifsnider, K. Some fundamental aspects of the fatigue and fracture response of composite materials. In Proceedings of the Fourteenth Annual Meeting of the Society of Engineering Science, Bethlehem, Palestine, 14–16 November 1977.
12. American Bureau of Shipping. *Rules for Building and Classing High-Speed Craft*; American Bureau of Shipping: Houston, TX, USA, 2015.
13. Suárez, J.C. The effect of slamming impact on out-of-autoclave cured prepregs of GFRP composite panels for hulls. *Proc. Eng.* **2016**, *167*, 252–264. [CrossRef]
14. Townsend, P.; Suárez, J.C. Reduction of slamming damage in the hull of high-speed crafts manufactured from composite materials using viscoelastic layers. *Ocean Eng.* **2018**, *159*. [CrossRef]

Article

Experimental Characterization of a High-Damping Viscoelastic Material Enclosed in Carbon Fiber Reinforced Polymer Components

Marco Troncossi [1,*], Sara Taddia [2], Alessandro Rivola [1] and Alberto Martini [1]

1 DIN–Dept. of Industrial Engineering, University of Bologna, Viale del Risorgimento 2, 40136 Bologna, Italy; alessandro.rivola@unibo.it (A.R.); alberto.martini6@unibo.it (A.M.)

2 BUCCI COMPOSITES S.p.A., via Mengolina 2, 48018 Faenza, Italy; s.taddia@bucci-industries.com

* Correspondence: marco.troncossi@unibo.it

Received: 8 August 2020; Accepted: 3 September 2020; Published: 6 September 2020

Abstract: This work aims to identify the damping properties of a commercial viscoelastic material that can be embedded and cured between the layers of composite laminates. The material may be adopted for reducing the vibration response of composite panels, typically used in automotive and aerospace applications, e.g., as vehicle body shell components. In order to objectively estimate the actual potential to enhance the noise vibration and harshness aspects, the effects of the viscoelastic material on the modal parameters of carbon/epoxy thin panels are quantitatively assessed through experimental modal analysis. Two different experiments are conducted, namely impact hammer tests and shaker excitation measurements. Based on the results of the experimental campaign, the investigated material is confirmed as a promising solution for possibly reducing the severity of vibrations in composite panels, thanks to its high damping properties. Indeed, the presence of just one layer proves to triple the damping properties of a thin panel. An approximate damping model is derived from the measured data in order to effectively simulate the dynamic response of new design solutions, including thin composite panels featuring the viscoelastic material.

Keywords: CFRP laminate; thin composite panel; viscoelastic material; vibration response; damping; experimental modal analysis

1. Introduction

The number of technological applications featuring composite materials is constantly increasing. For instance, composites are commonly adopted for manufacturing ship hulls and possible solutions to improve their resistance to slamming damage are currently investigated [1]. Additional examples are machine tools with parallel kinematics architecture [2] and robotic exoskeletons [3], to which the use of composite materials has been recently extended.

With reference to automotive applications, carbon fiber reinforced polymer (CFRP) laminates have been used in the past three decades to manufacture body shell components of sports cars, like the hood, since they exhibit a convenient combination of high strength and low weight [4]. Currently, they are under consideration also for possible use in other (mass-market) classes of cars, notwithstanding higher production costs [5,6]. Indeed, the use of lightweight materials is essential to lower the power consumption, particularly for hybrid and electric vehicles, hence, permitting them to fulfill sustainable mobility policies [4,5,7–9]. Moreover, they may permit to meet more stringent safety requirements in case of impacts [10–13].

The vehicle body shell components can be particularly critical from the noise vibration and harshness (NVH) standpoint. Indeed, such parts may exhibit undesired elastodynamic effects triggered by several sources, e.g., the engine/driveline operation, vibrations due to the road roughness, aerodynamic loads, and possible impacts with small particles, like raindrops [14–17]. This may result in a reduced comfort for driver and passengers. Hence, in this perspective, the dynamic response of CFRP panels for automotive applications must be carefully assessed.

Numerous advanced materials have been developed to limit the vibration severity of structures and components by increasing their structural damping [18–24]. While the research is still ongoing, promising results have been obtained [1,25–27]. In particular, Townsend et al. [1] studied the use of acrylonitrile butadiene styrene (ABS) cells filled with thermoplastic polyurethane (TPU) elastomer to improve the durability of glass fiber reinforced polymer (GFRP) hulls for high-speed boats. Liao et al. [26] investigated the influence of fiber orientation on the damping properties of CFRP cantilever beams featuring a single viscoelastic layer (copolymer of ethylene and acrylic acid—(PEAA)). Similarly, Berthelot and Sefrani [27] studied the effects of multiple design parameters (e.g., thickness and position of the viscoelastic layers) on the structural damping of GFRP cantilever beams with the addition of neoprene. Araùjo et al. [25] investigated the possibility of enhancing the structural damping of laminates with a viscoelastic core by optimizing the core thickness through genetic algorithms. This study focuses on a viscoelastic material that can be embedded and cured inside CFRP laminates, known as SMACWRAP®, manufactured by SMAC-Montblanc Technologies (Toulon, France). Such material has been preliminarily selected by the industrial partner involved in the research, on the basis of market analysis, supply chain policy, and technical aspects (e.g., stacking and lamination processes). Its effects on the modal parameters of thin carbon/epoxy panels are investigated by means of back-to-back experimental tests on very simple specimens. The main goal is to quantitatively assess its actual damping properties. To the authors best knowledge, whereas the dynamic response of composite sandwich beams with lightweight honeycomb core and SMACWRAP® layers has been assessed in [28,29], no data are available on thin panels.

The paper outline is as follows: Section 2 describes the tested specimens, the experimental setup and the signal processing techniques adopted for the analyses; a numerical model implemented to support the analysis of the data is also illustrated. In Section 3, the results of the experiments are reported and discussed; in addition, a possible approximate damping model is defined. In the final section, the main conclusions of the study are drawn.

2. Materials and Methods

2.1. Specimens

Two specimens are produced for the experimental tests, namely CFRP rectangular panels with a layer of SMACWRAP® embedded (referred to as *PY*) and without the viscoelastic material (*PN*), respectively.

The panel *PN* is a carbon/epoxy laminate consisting of four *twill* plies, hand lay-up with a *0°/0°/0°/0°* stacking sequence. Lamination is performed by means of vacuum bagging and autoclave process. The overall dimensions of the panel *PN* after lamination, namely the sides of the rectangular shape, *pl* and *ps*, and the total thickness, as well as its final mass are reported in Table 1.

Table 1. Specimen overall properties.

Specimen	Description	Long Side (*pl*) [mm]	Short Side (*ps*) [mm]	Thickness [mm]	Mass [kg]
PN	4 twill plies	800	580	1.49	1.008
PY	4 twill plies 3 SMACWRAP strips	800	580	1.62	1.055

The specimen *PY* has the same length (*pl*) and width (*ps*) of the panel *PN* (Figure 1, Table 1). It is produced by embedding three strips of SMACWRAP® of length *sl* and width *sw* (Table 2) between the two plies of the stack that are closest to the mold, according to the schematic shown in Figure 1, which represents the top view of the rectangular panel. Hence, the three strips lie on the same layer of the stack. A proper spacing, *cs*, is left between adjacent strips to improve the structural properties of the laminate, as suggested by the manufacturer of the viscoelastic material. Conversely, only a small spacing, *ss*, (or no spacing at all) is left between the strips and the panel sides, because the specimen is going to be clamped at all its edges for the experimental tests (see Section 2.2). Then, the panel is laminated with the same process adopted for the *PN* specimen. The geometric and mass properties of the *PY* panel, after lamination, are shown in Table 1.

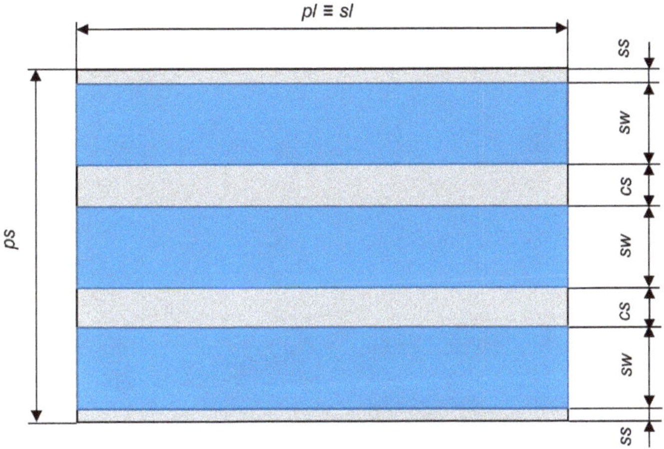

Figure 1. Schematic of the *PY* panel layout (top view).

Table 2. Properties of the SMACWRAP® strips embedded in the panel.

Parameter	Value
Density [kg/m^3]	1190
Thickness (single strip, uncured) [mm]	0.20
Strip length (*sl*) [mm]	800
Strip width (*sw*) [mm]	150
Side spacing (*ss*) [mm]	15
Central spacing (*cs*) [mm]	50

2.2. Experimental Setup and Test Procedures

In order to perform the measurements, each panel is fastened between two steel frames (Figure 2) by means of sixteen M10 through-screws, tightened with a controlled torque of 30 Nm. The size and mass properties of the steel frames are summarized in Table 3. The outer dimensions of the frames match the dimensions of the specimens. Hence, each specimen can vibrate as a rectangular plate characterized by the inner frame dimensions (*il* and *is* in Figure 2a and Table 3) and clamped at its edges. This kind of constraint aims to simulate typical boundary conditions featured by CFRP panels used in automotive applications, such as the car roof fixed to the vehicle body. Then, the steel frame is suspended by using low-stiffness elastic ropes, with the short sides vertically directed (Figure 2b).

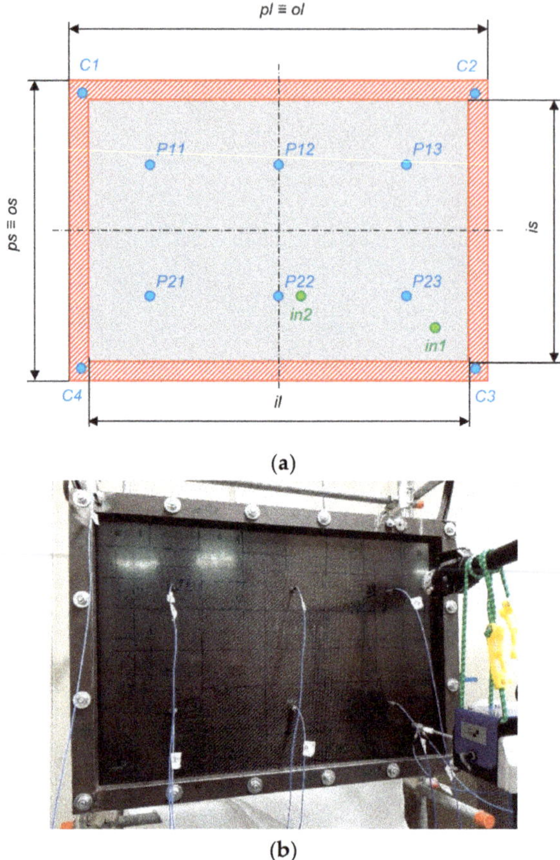

Figure 2. Experimental setup: (**a**) schematic of the steel frame and of the excitation/measurement nodes; (**b**) photo of the specimen *PN* excited by the shaker at node *in1*.

Table 3. Properties of the steel frames.

Parameter	Value
Total mass (2 frames + bolts) [kg]	11.54
Outer long side (*ol*) [mm]	800
Outer short side (*os*) [mm]	580
Inner long side (*il*) [mm]	720
Inner short side (*is*) [mm]	500

Experimental modal analysis (EMA) is performed by using two different excitation techniques, namely impact and shaker excitations. A modal geometry consisting of 12 nodes is adopted to represent the specimens (Figure 3). In particular, 6 nodes (referred to as Pij, $i = 1, 2$ and $j = 1, \ldots, 3$) are associated with the panel; 4 nodes (Ck, $k = 1, \ldots, 4$) are associated with the corners of the steel frame; 2 nodes (*in1* and *in2*) are the locations on the panels excited by the shaker. It can be noticed that only six accelerometers are attached to the panels, to not significantly alter the tested system dynamics due to excessive additional masses. Such a limited spatial resolution is reasonably supposed to cause an aliased estimation of the mode shapes. A simple finite element (FE) model of the panels was thus preliminarily implemented with the aim (*i*) to let the numerical modal analysis support the experimental data processing (see Section 2.4) and (*ii*) to pinpoint adequate locations of nodes Pij

for a straightforward visual identification of the main mode shapes. The vibration response of the specimens along the Z direction (perpendicular to the panel plane) is measured at all nodes *Pij* and at node *C1* by means of IEPE piezoelectric accelerometers (PCB Piezotronics Inc., Depew, NY, USA). Node *C1* permits to monitor the behavior of the steel frame, in order to possibly identify spurious mode shapes involving vibrations of the frame itself, which should be excluded from the analysis.

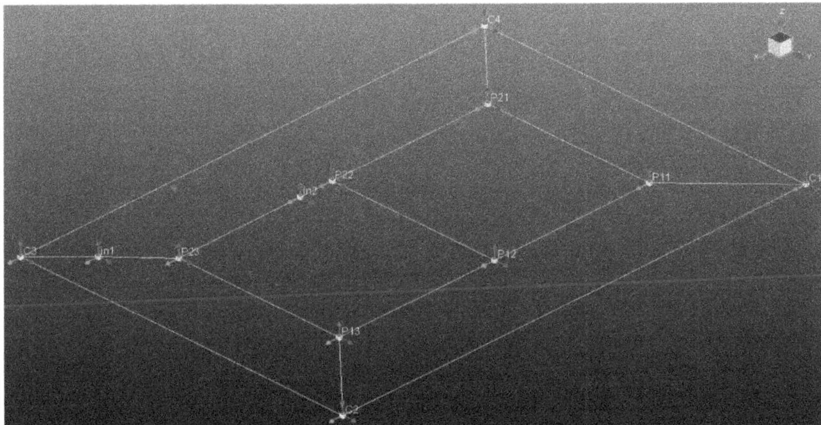

Figure 3. Geometry of the modal model: *C2–C4* are active in the impact tests only; *in1* and *in2* are active in the shaker tests only.

Impact test is performed by exciting all the specimens' nodes except *in1* and *in2*, perpendicularly to the panel plane, with an impulse force hammer (PCB 086C04, PCB Piezotronics Inc., Depew, NY, USA). Hence, a total amount of 7 response nodes and 10 reference nodes (i.e., the nodes impacted by the hammer) are used for the analysis (*in1* and *in2* are deactivated in the modal model). This setup may be seen as the combination of the so-called *roving hammer* and *roving sensors* approaches, thus generating a redundant dataset that may be exploited to find the most convenient subset of frequency response functions (FRFs) estimates that EMA is based on.

The second test is conducted by exciting the specimens with an electrodynamic shaker (TMS-K2007E01, The Modal Shop Inc., Cincinnati, OH, USA) along the Z direction at the input nodes *in1* and *in2* (Figure 3). The excitation force and the acceleration at nodes *in1* and *in2* are measured by an IEPE piezoelectric impedance head (PCB 288D01, PCB Piezotronics Inc., Depew, NY, USA). The input nodes *in1* and *in2* are excited in distinct runs (i.e., not simultaneously), so that only a reference node at a time is considered in the analyses. The results obtained by using the *in2* reference point basically match those obtained with excitation applied at node *in1*. Hence, only the results concerning the *in1*-case are reported in this paper, whereas the *in2*-case is neglected hereafter. For a better graphical representation of the results, nodes *C2*, *C3*, and *C4* are kept visible in the modal geometry also in shaker tests, although no data are associated with them. Two different excitation profiles are adopted, namely burst random and chirp signals, and two excitation levels are set for each one (also to investigate possible non-linear dynamic response of the CFRP panels).

All the experimental campaigns are performed by using an LMS SCADAS SCM-05 frontend and LMS Test.Lab software package (Siemens Digital Industries Software, Plano, TX, USA), with the following acquisition setup parameters:

- Sampling frequency: F_{Sh} = 512 Hz (impact hammer tests), F_{Ss} = 2048 Hz (shaker tests);
- Bandwidth: BW_h = 0.5–256 Hz (impact hammer tests), BW_s = 5–1024 Hz (shaker tests);
- Acquisition duration: T = 8 s (spectral frequency resolution: Δf = 0.125 Hz);
- Number of averages: N_{av} = 10;

- Shaker excitation levels: L_R = 0.2 V, 1 V (Random tests), L_C = 0.1 V, 0.5 V (Chirp tests).

The frequency bandwidth in impact tests is lower than in shaker tests due to practical constraints: the excitation is in fact provided by an impact hammer featuring a plastic covered tip (to preserve the thin panels integrity), and the corresponding force spectrum proves almost flat up to 300 Hz only.

2.3. Signal Processing and Analysis

It is known that structural damping is generally difficult to identify and even more difficult to model. The literature offers a number of parameters to quantify damping, depending on the specific sector of application, e.g., *damping ratio*, *logarithmic decrement*, *loss factor*, *quality factor*, *decay constant*. In practice, they carry very similar information, being interrelated through analytical formulations [30]. In the field of sound and vibration, the damping properties are commonly determined in the Frequency domain. In particular, from the EMA/modal identification, modal coefficients related to the several identified mode shapes are obtained; from these ones, a damping ratio per mode shape can be defined. Indeed, in this study, the damping ratio (ζ) is chosen to quantitatively assess the structural damping associated with each vibration mode.

For all the tests, the FRFs between response and reference signals, acceleration and force respectively, are estimated. Then, the modal parameters of the specimens, i.e., mode shapes X_i, natural frequencies f_i, and damping ratios ζ_i ($i = 1, 2 \ldots$) are computed.

Numerous data processing are performed for each of the five experimental campaign (impact test and four shaker tests), in order to get reliable and robust results. Indeed, it is well known that EMA results might prove significantly sensitive to the frequency band and/or the FRF set chosen for the analysis, as well as to the values set up for certain parameters of EMA algorithms. In particular: algorithms in the time and frequency domains are used; different combinations of FRF sets are selected as the basis for the modal parameter computations; different bandwidths are defined (e.g., 5–1000 Hz, 5–200 Hz, 200–1000 Hz).

2.4. FE Model

The *PN* panel is modelled by using the FE analysis software *Ansys*® (Ansys Inc., Canonsburg, PA, USA). After a—straightforward—mesh convergence analysis, a regular mesh of 630 nodes is generated by using 4-noded quadrilateral shell elements (*SHELL181*). The 6 Degrees of Freedom (DOFs) of all the nodes on the edges of the panel are fixed to the ground to represent (ideal) clamped boundary conditions. The composite material is modelled as a homogeneous equivalent material, by assigning laminate material properties. This simplified representation is assumed as largely adequate to evaluate the global dynamic response of the panel [31]. Indeed, the numerical modal analysis of the clamped plate is performed to support the correct interpretation of mode shapes retrieved from EMA avoiding the possible spatial aliasing issues mentioned in Section 2.2. The accurate estimation of the natural frequencies is not required; hence, model updating to match the experimental resonances is not performed in this study.

3. Results and Discussion

The redundancy of both the test campaigns and the analyses prove fundamental for the final outcomes, as the correct estimation of the modal parameters has been possible only by exploiting different combinations. As an example, Figure 4 reports the FRF-sum amplitudes obtained from burst random (excitation level 1 V) and Chirp periodic (excitation level 0.5 V) tests, respectively. The former is globally less scattered, thus making the response peaks emerge very clearly. On the other hand, the latter can catch resonances of modes that are poorly excited in the previous case.

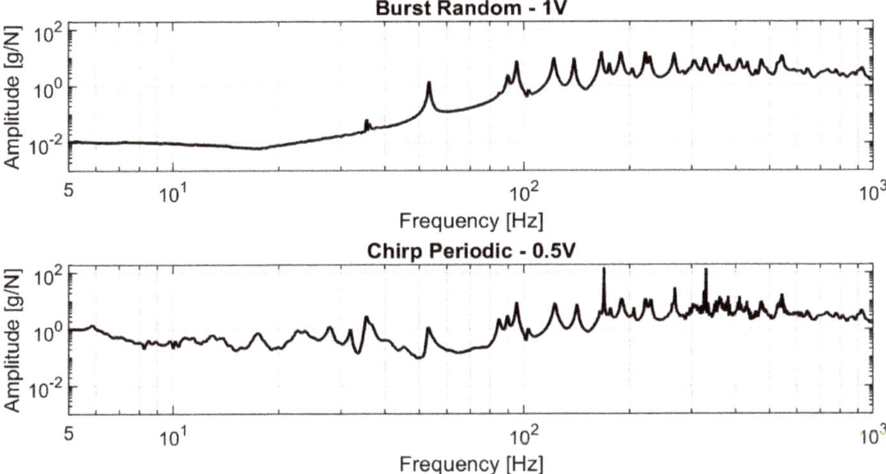

Figure 4. Sum of frequency response functions (FRFs), limited to 5–1000 Hz, estimated in two tests performed on panel *PN*.

3.1. Impact Hammer Tests

The impact hammer testing (practically simpler to set up) proves inadequate to perform the EMA correctly: indeed, in this case, acceptable values are hardly obtained for the three main parameters that indicate the reliability of the EMA results, namely modal phase collinearity (MPC), mean phase deviation (MPD), and modal participation (MP) [32,33]. Overall results are not consistent with the modal parameters computed through numerical analysis nor with those estimated through shaker tests EMA, and thus they are not reported in the following. This trouble might be reasonably explained considering the effects of localized impact forces on thin CFRP panels: significant local deformations may prevent the panel mode shapes to be effectively excited. The possibility of achieving more satisfactory results by using a miniature impact hammer may be verified in future tests.

3.2. Shaker Excitation Tests

The results obtained from the analyses of the shaker excitation measurements for the first ten vibration modes are summarized in Table 4, where subscripts *PN*, *PY*, and *i* refer to panel *PN*, panel *PY*, and mode ranking, respectively. Such results are computed as an average of the values obtained from the different modal identification processes, by also varying the excitation profile/level and the computational algorithm (in the time and frequency domains, respectively), while keeping fixed the frequency range (5–200 Hz). Only the results characterized by satisfactory values of the parameters MPC, MPD, and MP have been included in the average. The mode shapes are described by means of two parameters, namely m and n, according to a conventional notation frequently adopted for the vibrations of rectangular plates [34]. In particular, in the case of a rectangular plate made of a homogeneous material with all the sides clamped, m and n refer to the number of half-sine waves along the direction of the long and short sides, respectively. The comparison between the natural frequencies and the damping ratios of the *PN* and *PY* panels is also reported, in terms of percentage variations Δf_i and $\Delta\zeta_i$, respectively, normalized to the panel PN values; moreover, factors of damping increment (i.e., the ratios between the damping ratios) $\rho_{YN,i} = \zeta_{PY,i}/\zeta_{PN,i}$ are provided as well. The result presentation is limited to the first ten modes, below 200 Hz, since, for higher ranking modes—above all for panel *PY*—the mode shape estimation proves not extremely reliable (with fuzzy values for MPC, MPD, and MP parameters) and a robust association with the *PN* panel ones is not straightforward, thus making the direct comparison uncertain.

Table 4. Modal parameters estimated from shaker tests experimental modal analysis (EMA).

Mode Ranking		Mode Shape	Natural Frequency		Damping Ratio		Comparison		
$\#_{PN,i}$	$\#_{PY,i}$	(Xi)	$f_{PN,i}$ [Hz]	$f_{PY,i}$ [Hz]	$\zeta_{PN,i}$ [%]	$\zeta_{PY,i}$ [%]	Δf_i [Hz]	$\Delta\zeta_i$ [%]	$\rho_{YN,i}$ [−]
1	1	m = 1, n = 1	35.2	34.3	0.36	1.45	−2.7	302.8	4.0
2	2	m = 2, n = 1	53.0	48.3	0.50	1.46	−8.9	192.0	2.9
3	4	m = 1, n = 2	83.3	85.4	0.49	1.66	2.5	238.8	3.4
4	3	m = 3, n = 1	90.1	83.1	0.65	1.67	−7.8	156.9	2.6
5	5	m = 2, n = 2	95.1	104.2	0.52	1.16	9.7	123.1	2.2
6	6	m = 3, n = 2	121.1	118.3	0.66	1.69	−2.3	156.1	2.6
7	7	m = 4, n = 1	136.9	127.9	0.52	1.81	−6.6	248.1	3.5
8	8	m = 4, n = 2	163.3	160.8	0.74	1.67	−1.5	125.7	2.3
9	9	m = 1, n = 3	164.3	175.1	0.46	1.67	6.5	263.0	3.6
10	10	m = 2, n = 3	174.9	189.6	0.57	1.88	8.4	229.8	3.3

The comparison between the FE model (FEM) and the EMA results for the *PN* panel vibration modes #1, #6, and #9 (chosen as examples) is reported in Figure 5.

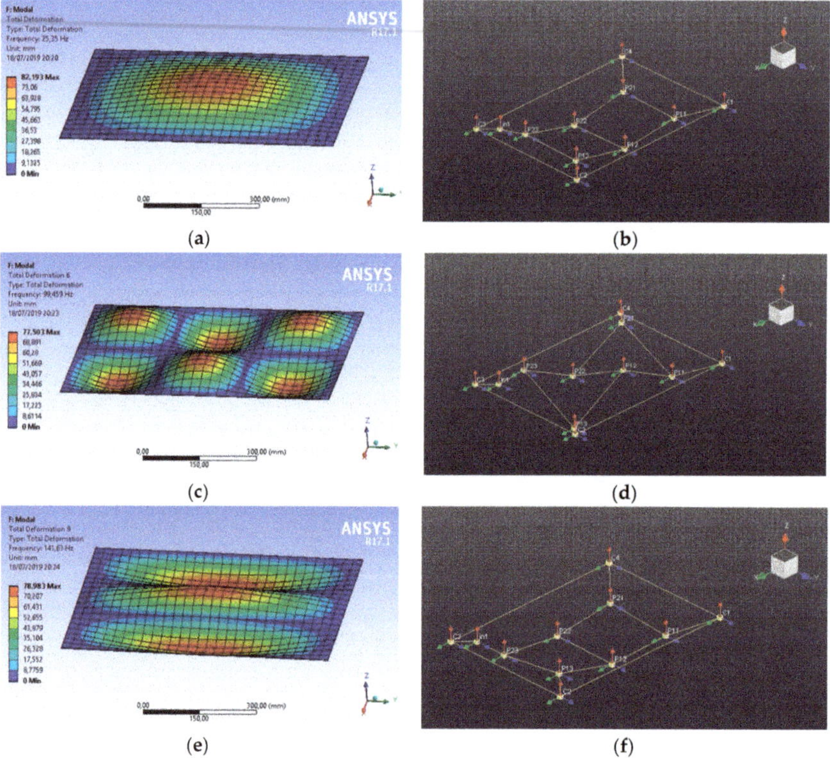

Figure 5. *PN* panel results, finite element model (FEM) vs. EMA vibration modes: (**a**) FEM #1; (**b**) EMA #1; (**c**) FEM #6; (**d**) EMA #6; (**e**) FEM #9; (**f**) EMA #9.

In particular, it can be observed that the limited resolution of the modal geometry (*Pij*, *C1*, and *in1* nodes) causes a very similar representation for the mode shapes #1 and #9. This confirm the usefulness of the FEM results for properly identifying the mode shapes estimated through the EMA, thus, partially solving the shape aliasing problem.

The natural frequency variations between the two panels remain rather limited for all the vibration modes, the maximum difference being less than 10%. The mode ranking remains the same except for modes #3 and #4. The trend of Δf_i is not monotonic: in particular $\Delta f_i > 0$ for modes with $n \geq m$ (apart from $n = 1$), due to the stiffening effect of the additional thickness of the SMACWRAP® layer on the mode shapes featuring more waves along the short side with respect to the long one (e.g., mode #9 in Figure 5e). On the contrary, the slight decrement in other natural frequencies is due to the additional mass effect.

The damping ratios characterizing the *PY* panel are significantly higher than in the *PN* panel, for all the vibration modes, with percentage increment ranging from 123% to 303%, depending—apparently inconsistently—on the single modes. For an overall evaluation, Table 5 reports the mean values (μ), standard deviations (σ) and relative standard deviations ($\sigma^* = \sigma/\mu$ 100) of the damping ratios computed for the two panels and their fraction (i.e. the results shown in Table 4). On the average, the presence of the viscoelastic material triples the damping properties of the panel, thus, proving its effectiveness as a possible solution to dampen vibrations in thin CFRP panels. The measured damping ratios are basically in agreement with the values found in [28] for sandwich beams with lightweight honeycomb core and one SMACWRAP® layer. However, therein a smaller increment in the damping properties is generally observed, since the untreated specimen exhibits higher damping ratios.

Table 5. Statistical parameter values of the EMA damping ratios and corresponding fraction.

Parameter	ζ_{PN}	ζ_{PY}	ρ_{YN}
Mean value, μ	0.55	1.61	3.04
Standard deviation, σ	0.11	0.20	0.59
Relative standard deviation, σ^*	19.2%	12.1%	19.3%

Further indications on the effectiveness of the tested viscoelastic material may be obtained from a comparison with the results presented by Liao et al. [26] and by Berthelot and Sefrani [27]. In the former, the damping ratio of the first mode of a CFRP cantilever beam (thickness of 1.5 mm) appears to be almost doubled by adding a 0.1 mm layer of PEAA. In the latter, one result is that an increment of almost three times could be achieved for the first mode of a GFRP cantilever beam (thickness of 2.4 mm) with a 0.2 mm layer of neoprene.

3.3. Damping Model of the Specimens

The vibration modes identified through the analyses of the shaker excitation measurements in the frequency range 5–1000 Hz are 39 and 21 for the *PN* and *PY* panels, respectively. The damping ratio experimental estimates are plotted over the corresponding natural frequencies in Figure 6.

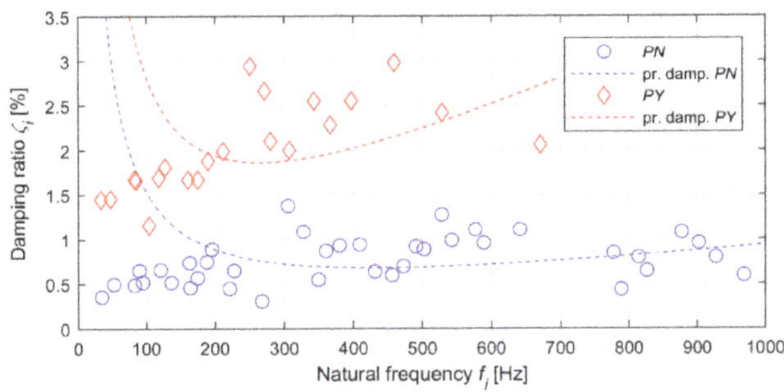

Figure 6. Modal parameters with proportional damping interpolating curves.

Firstly, data fitting by means of a proportional damping model is evaluated. With this popular model, also known as Rayleigh model [35,36], solving the structural dynamics/multibody dynamics problem is remarkably simpler. Indeed, it is largely adopted (e.g., for FE modelling and analysis), even if it may lead to inaccurate results [37]. Under the hypothesis of proportional damping, the following relation holds:

$$2\zeta_i\omega_i = \frac{c_i}{m_i} = \frac{\alpha m_i + \beta k_i}{m_i} = \alpha + \beta\omega_i^2 \tag{1}$$

where α and β are the proportional damping coefficients, $\omega_i = 2\pi f_i$, and m_i, k_i, c_i, are the modal mass, modal stiffness, and modal damping of the i-th mode, respectively. After determining the modal parameters f_i and ζ_i for N vibration modes, the coefficients α and β can be estimated through least-squares approximation, by using the following expression:

$$\left\{ \begin{array}{c} \alpha \\ \beta \end{array} \right\} = \left(\mathbf{A}^T\mathbf{A}\right)^{-1}\mathbf{A}^T\mathbf{b} \tag{2}$$

where

$$\mathbf{A} = \begin{bmatrix} 1 & \omega_1^2 \\ \vdots & \vdots \\ \vdots & \vdots \\ 1 & \omega_N^2 \end{bmatrix};\ \mathbf{b} = 2\left\{ \begin{array}{c} \zeta_1\omega_1 \\ \vdots \\ \vdots \\ \zeta_N\omega_N \end{array} \right\}. \tag{3}$$

The proportional damping curves computed for the *PN* and *PY* panels are plotted in Figure 6. From the comparison with the corresponding experimental data, such models do not appear suitable for the tested specimens, since the damping ratios of modes at the lowest natural frequencies are not correctly fitted.

The EMA results show that the *PY* panel exhibits a quite evident growth of the damping ratios as the natural frequencies increase. A slight increment in ζ_i apparently characterizes the *PN* panel as well. Linear regression models, described by the expression

$$\zeta_i = \lambda_0 + \lambda_1 f_i, \tag{4}$$

are, thus, computed for possibly describing the observed trends of the damping ratios as linear functions of the natural frequencies. The experimental values of the damping ratios and the corresponding regression lines are plotted in Figure 7 for both the panels. The estimates of the intercept (λ_0) and of the slope (λ_1) of the regression lines are reported in Table 6, with the corresponding p-values. It is worth recalling that the p-value measures the probability that the estimated parameter has a correlation with the measured data, i.e., whether it is relevant for the regression model (a p-value of 0.05 or lower being the level of significance commonly accepted to reject the null hypothesis). For each specimen, the coefficient of determination, R^2 (also known as *R-squared*), and the p-value of the regression model as a whole are also shown. The former, ranging from 0 to 1, indicates which fraction of the variability exhibited by the dependent variable is explained by the regression model. The latter assesses the level of significance of the regression model and, for the models here considered, basically coincides with the *p-value* of λ_1, since there is only one independent variable (i.e., the frequency).

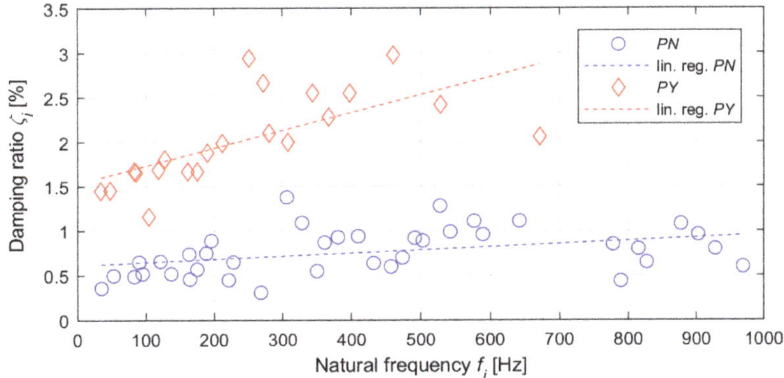

Figure 7. Modal parameters with linear regression fitting lines.

Table 6. Linear regression parameters.

Specimen	Coefficient	Estimate	Coefficient p-Value	R^2	Model p-Value
PN	λ_0 λ_1	0.613 3.49×10^{-4}	1.82×10^{-10} 1.73×10^{-2}	0.144	1.73×10^{-2}
PY	λ_0 λ_1	1.54 1.99×10^{-3}	4.48×10^{-9} 9.81×10^{-4}	0.444	9.81×10^{-4}

The regression model explains almost 45% of the variability in the damping ratios of the *PY* panel ($R^2 = 0.444$) and appears sufficiently reliable (p-value $= 9.81 \times 10^{-4}$). Hence, it can be considered as a satisfactory approximation of the *PY* damping trend. Regarding the *PN* panel, the regression model accounts for only 15% of the data variability, but the approximation still appears acceptable. Additionally, the linear regression model clearly appears to approximate better the data, when compared to the Rayleigh damping hypothesis. The comparison between the two computed regression lines further confirms the dampening capabilities of the tested viscoelastic material, over a quite wide frequency range.

It is worth noting that the regression model of the *PY* panel represents the behavior of the composite laminate and not the intrinsic characteristics of the viscoelastic material. The usefulness of its implementation may be thus limited, but identifying the damping properties of only a single viscoelastic layer through EMA (which this study is based on) is objectively impossible, due to its extremely low thickness. Nonetheless, it may be used as a reference for predicting the dynamic response of CFRP laminates featuring overall characteristics similar to the panels targeted in this study (which are assumed reasonably quite common in a number of applications). For instance, it may be adopted to perform structural dynamics analyses of CFRP laminates within a FE code, in case the composite can be modelled with an equivalent material (see Section 2.4). Therefore, the proposed damping model may usefully support the investigation of new design solutions of components that include thin composite panels featuring the investigated viscoelastic material: in particular, when the target of elastodynamic simulations is the back-to-back comparison of different design solutions, the absolute accuracy of results is often of secondary importance.

4. Conclusions

This study investigates, through experimental modal analysis, the damping properties of a commercial viscoelastic material that can be embedded in CFRP laminates for reducing the vibration response. Impact tests and shaker excitation tests are performed on thin panels in clamped boundary conditions. The former proves not adequate to provide satisfactory results, due to the local deformation of the thin panels under impact loads. The presented results, retrieved from many combinations of signal analyses performed on shaker test data, show a significant increment of the damping ratio over a wide frequency range. Hence, the viscoelastic material proves its high damping capability and is thus confirmed as a promising solution to be possibly adopted in automotive applications for addressing NVH issues.

Finally, a simple damping model is formulated based on linear regression of the measured data. The proposed model may be adopted for a preliminary estimation of the dynamic response of carbon/epoxy thin panels featuring the investigated viscoelastic material, thus possibly guiding the design process of new composite structure solutions in the early phases.

Author Contributions: Conceptualization, M.T. and S.T.; Methodology, M.T.; Software, M.T. and S.T.; Validation, M.T.; Formal Analysis, M.T. and A.M.; Investigation, M.T. and A.M.; Resources, M.T., S.T. and A.R.; Data Curation, M.T. and S.T.; Writing—Original Draft Preparation, M.T. and A.M.; Writing—Review and Editing, M.T., S.T., A.R. and A.M.; Visualization, M.T. and A.M.; Supervision, M.T.; Project Administration, M.T. All authors have read and agreed to the published version of the manuscript.

Funding: This research received no external funding.

Conflicts of Interest: The authors declare no conflict of interest.

References

1. Townsend, P.; Suárez, J.C.; Sanz-Horcajo, E.; Pinilla-Cea, P. Reduction of slamming damage in the hull of high-speed crafts manufactured from composite materials using viscoelastic layers. *Ocean Eng.* **2018**, *159*, 253–267. [CrossRef]
2. Neumann, K.-E. *True Mobile/Portable Drilling and Machining, a Paradigm Shift in Manufacturing*; SAE Technical Paper; SAE International: Warrendale, PA, USA, 2017. [CrossRef]
3. Pappas, G.A.; Botsis, J. Design optimization of a CFRP–aluminum joint for a bioengineering application. *Des. Sci.* **2019**, *5*, 14. [CrossRef]
4. Fuchs, E.R.; Field, F.R.; Roth, R.; Kirchain, R. Strategic materials selection in the automobile body: Economic opportunities for polymer composite design. *Compos. Sci. Technol.* **2008**, *68*, 1989–2002. [CrossRef]
5. Lutsey, N. *Review of Technical Literature and Trends Related to Automobile Mass-Reduction Technology*; Research Report—UCD-ITS-RR-10-10; Institute of Transportation Studies, University of California: Davis, CA, USA, 2010.
6. Anwar, M.; Sukmaji, I.C.; Wijang, W.R.; Diharjo, K. Application of carbon fiber-based composite for electric vehicle. *Adv. Mater. Res.* **2014**, *896*, 574–577. [CrossRef]
7. Arifurrahman, F.; Budiman, B.A.; Aziz, M. On the lightweight structural design for electric road and railway vehicles using fiber reinforced polymer composites—A Review. *Int. J. Sustain. Transp. Technol.* **2018**, *1*, 21–29. [CrossRef]
8. Tsirogiannis, C.E. Design and modelling methodologies of an efficient and lightweight carbon-fiber reinforced epoxy monocoque chassis, suitable for an electric car. *Mater. Sci. Eng. Adv. Res.* **2017**, *2*, 5–12. [CrossRef]
9. Odabaşı, V.; Maglio, S.; Martini, A.; Sorrentino, S. Static stress analysis of suspension systems for a solar-powered car. *FME Trans.* **2019**, *47*, 70–75. [CrossRef]
10. Sukmaji, I.C.; Anwar, M.; Wijang, W.R.; Danardono, D.P.D. Hybrid carbon-glass fiber composite for the door electric car application. In Proceedings of the 2013 Joint International Conference on Rural Information & Communication Technology and Electric-Vehicle Technology (rICT & ICeV-T), Bandung, Indonesia, 26–28 November 2013; Institute of Electrical and Electronics Engineers (IEEE): Piscataway, NJ, USA, 2013; pp. 1–3.
11. Sudirja Hapid, A.; Kaleg, S.; Budiman, A.C.; Amin. The crumple zone quality enhancement of electric cars bumper fascia using a carbon fiber reinforced vinyl ester—Microsphere composites. In Proceedings of the 2019 International Conference on Sustainable Energy Engineering and Application (ICSEEA), Tangerang, Indonesia, 23–24 October 2019.
12. Ahmed, A.; Wei, L. Introducing CFRP as an alternative material for engine hood to achieve better pedestrian safety using finite element modeling. *Thin Walled Struct.* **2016**, *99*, 97–108. [CrossRef]
13. Bang, S.; Park, Y.; Kim, Y.; Shin, T.; Back, J.; Lee, S.K. Effect of the fiber lamination angle of a carbon-fiber, laminated composite plate roof on the car interior noise. *Int. J. Automot. Technol.* **2019**, *20*, 73–85. [CrossRef]

14. Raghuvanshi, J.; Palsule, A.; Bodhale, N.; Kharade, A.; Pol, A. *Sensitivity Study of Different Damping Treatments Using Simulation and Physical Testing Methodologies on Structure Borne Driver's Ear Noise Performance in a Premium Hatchback Car*; SAE Technical Paper; SAE International: Warrendale, PA, USA, 2019. [CrossRef]
15. Gur, Y.; Wagner, D. Damping properties and NVH modal analysis results of carbon fiber composite vehicle components. *SAE Int. J. Mater. Manuf.* **2017**, *10*, 198–205. [CrossRef]
16. Yu, Z.; Cheng, D.; Huang, X. Low-frequency road noise of electric vehicles based on measured road surface morphology. *World Electr. Veh. J.* **2019**, *10*, 33. [CrossRef]
17. Martini, A.; Bellani, G.; Fragassa, C. Numerical assessment of a new hydro-pneumatic suspension system for motorcycles. *Int. J. Automot. Mech. Eng.* **2018**, *15*, 5308–5325. [CrossRef]
18. House, J.R.; Hilliar, A.E. Vibration Damping Materials. Patent WO-90/01645, 22 February 1990.
19. Fujimoto, J.; Tamura, T.; Furihata, T.; Suzuki, Y.; Kauchi, K. Laminated Vibration-Damping Material. U.S. Patent US-005368916, 29 November 1994.
20. Sutton, S.P.; Principe, F.; Gentile, M.M. Vibration Damping Composite Material. U.S. Patent US-5965249, 12 October 1999.
21. Ellis, J.; Hadley, P. Composite Material. Patent WO-2014/147243, 25 September 2014.
22. Sumita, M.; Kaneko, H.; Murase, K. Composite Damping Material. U.S. Patent US-20160040744A1, 11 February 2016.
23. Stopin, G.; Tesse, C. Constrained-Layer Damping Material. U.S. Patent US-9243402B2, 26 January 2016.
24. Alexander, J.H.; Eichhorn, G.; Gerdes, R.W.; Hanschen, T.P.; Herdtle, T.; Yoo, T. Multilayer Damping Material. U.S. Patent US-20180156296A1, 7 June 2018.
25. Araújo, A.; Martins, P.; Soares, C.M.; Herskovits, J. Damping optimization of viscoelastic laminated sandwich composite structures. *Struct. Multidiscip. Optim.* **2009**, *39*, 569–579. [CrossRef]
26. Liao, F.-S.; Su, A.-C.; Hsu, T.-C.J. Vibration damping of interleaved carbon fiber-epoxy composite beams. *J. Compos. Mater.* **1994**, *28*, 1840–1854. [CrossRef]
27. Berthelot, J.-M.; Sefrani, Y. Damping analysis of unidirectional glass fiber composites with interleaved viscoelastic layers: Experimental investigation and discussion. *J. Compos. Mater.* **2006**, *40*, 1911–1932. [CrossRef]
28. Fotsing, E.; Sola, M.; Ross, A.; Ruiz, E. Lightweight damping of composite sandwich beams: Experimental analysis. *J. Compos. Mater.* **2012**, *47*, 1501–1511. [CrossRef]
29. Piollet, E.; Fotsing, E.R.; Ross, A.; Michon, G. High damping and nonlinear vibration of sandwich beams with entangled cross-linked fibres as core material. *Compos. Part B Eng.* **2019**, *168*, 353–366. [CrossRef]
30. Gade, S.; Herlufsen, H. Digital filter vs fft techniques for damping measurements. *J. Sound Vib.* **1990**, *24*, 24–32.
31. Barbero, E.J. *Finite Element Analysis of Composite Materials Using Ansys®*; Informa UK Limited: London, UK, 2013.
32. Heylen, W.; Lammens, S.; Sas, P. *Modal Analysis Theory and Testing*, 2nd ed.; Katholieke Universiteit Leuven: Leuven, Belgium, 1998.
33. Juang, J.-N.; Pappa, R.S. An eigensystem realization algorithm for modal parameter identification and model reduction. *J. Guid. Control. Dyn.* **1985**, *8*, 620–627. [CrossRef]
34. Leissa, A.W. *Vibrations of Plates*; National Aeronautics and Space Administration (NASA): Washington, DC, USA, 1969.
35. Ewins, D.J. *Modal Testing: Theory, Practice and Application*, 2nd ed.; Research Studies Press Ltd.: Baldock, UK, 2000.
36. Rao, S.S. *Mechanical Vibrations*, 5th ed.; Prentice Hall: Upper Saddle River, NJ, USA, 2004.
37. Charney, F.A. Unintended consequences of modeling damping in structures. *J. Struct. Eng.* **2008**, *134*, 581–592. [CrossRef]

Article

Comparative Study of Glass Fiber Content Measurement Methods for Inspecting Fabrication Quality of Composite Ship Structures

Zhiqiang Han [1], Sookhyun Jeong [1], Jackyou Noh [2] and Daekyun Oh [3,*]

1 Department of Ocean System Engineering, Mokpo National Maritime University, Mokpo 58628, Korea; hzq910413@gmail.com (Z.H.); jeongsookhyun@gmail.com (S.J.)
2 Department of Naval Architecture & Ocean Engineering, Kunsan National University, Gunsan 54150, Korea; snucurl@kunsan.ac.kr
3 Department of Naval Architecture and Ocean Engineering, Mokpo National Maritime University, Mokpo 58628, Korea
* Correspondence: dkoh@mmu.ac.kr; Tel.: +82-61-240-7238

Received: 2 June 2020; Accepted: 24 July 2020; Published: 26 July 2020

Abstract: A comparative study of glass fiber content (Gc) measurement methods was conducted using actual glass fiber reinforced plastic laminates from the hull plate of a 26-ton yacht. Two prototype side hull plates with the design Gc (40 wt.%) and higher Gc (64 wt.%) were prepared. Four methods were used to study the samples: the calculation method suggested by classification societies' rules; two direct measurement methods using either calipers and scales or a hydrometer; and the burn-off method, wherein the resin matrix is combusted from the laminates. The results were compared and analyzed to identify the accuracy and benefits of each method. The rule calculation method was found to be effective if the quality of the manufacturing process is known. However, fabrication errors in the laminate structures cannot be detected. Additionally, while direct methods are used to measure the density of glass fibers using measurements of the densities of raw materials and laminates, the volume of inner defects occurring during the fabrication of laminates could not be considered. Finally, it was found that the burn-off method measures Gc and considers the defect volume (voids) inside laminates as well as the non-uniformity of the external shape.

Keywords: composite ship; composite structure; glass fiber content; void volume; burn-off test; calcination test

1. Introduction

Glass fiber reinforced plastics (GFRPs) have been widely used for decades for building small ships, such as fishing boats and yachts [1,2], as they exhibit good specific strength, corrosion resistance, and excellent workability. Many GFRP ships are manufactured with a large design margin, as this enables faster and cheaper production. However, it also results in heavier vessels due to the thicker GFRP laminate structures. Laminate structures for composite ships are considerably thicker than those used for aviation and automobile components, which can cause adverse structural effects such as fatigue over the ship's lifetime [3]. In addition, GFRPs have elicited environmental concerns related to their poor recyclability and issues during drying and disposal [4,5]. Accordingly, increased research attention has been directed toward the optimal design and weight reduction of GFRP ships.

The design regulations for GFRP structures of small crafts are addressed in the international standard ISO 12215-5 [6]. The laminate thickness, controlled by the number of glass fiber cloth layers (plies), is designed based on ship variables such as hull shape, displacement, speed, and hull form, as well as structural variables such as the layout of stiffeners, design loads, and the design of the composite material (type of reinforcement material, reinforcement method, and mechanical properties) [7,8]. The glass fiber weight fraction, Gc, is a critical design element that significantly affects the mechanical properties of the laminate. The ISO standards and classification society rules provide equations for estimating the mechanical properties of laminate structures according to variations in Gc. These theoretical equations are also differentiated according to the type of glass fiber, reinforcement method, number of plies, and amount of resin used [6,9,10].

However, while Gc is an important indicator of the fabrication quality of a laminate structure [11], the quality of the laminate can vary widely according to the manufacturing method, environment, and operator skill level. GFRP ships are typically manufactured using the hand lay-up method, which can lead to degradation of the glass fiber or resin due to human error [12,13]. Moreover, the laminate structures often contain fabrication defects, such as porosity or voids. These fabrication defects can have a significant impact on the physical performance of the laminate structure, even if Gc meets the design parameters [14–16].

For ships that adopt special glass fibers, special structures, or structures thinner than the rules allow, classification societies require manufacturers to disclose the Gc of the laminate structure along with the results of material tests. The mechanical properties are often verified by fracture testing according to the ASTM standards [3,7,11,17], while experimental methods (e.g., resin burn-off) or theoretical calculations are recommended to determine Gc. Nevertheless, the classification rules do not provide detailed specifications on determining Gc. In addition, because of the flexible fabrication characteristics and different types and combinations of materials, it is difficult to verify that laminates are fabricated according to the designed Gc.

To aid the design and manufacture of GFRP laminates for shipbuilding, it is critical to identify an accurate and consistent method of determining Gc. The method should also measure the size and volume of voids inside the laminate structure, as well as the measurement error of the outer shape. Considering the diverse methods currently in use to determine Gc and defect incidence, a comparative study of the different methods is of significant applicative interest. Herein, we empirically tested four different methods of measuring the Gc of GFRP ship components: a widely used theoretical calculation proposed in the classification rules; direct measurement of the volume and weight or relative density of a specimen followed by calculation of Gc; and the burn-off method, wherein the resin matrix is combusted from the laminate. The methods were used to assess two types of composite hull plates for a 16-m ship. A comparative analysis of the results methods was then conducted to determine:

1. The reliability of each method for analyzing the laminate structure and establishing the manufacturing precision according to the material design
2. The accuracy of each method for testing the fabrication quality of the laminate structure
3. The advantages and disadvantages of each method

Overall, we found that the calculation method is effective at measuring Gc if the quality of the manufacturing process is known, but that it cannot detect fabrication errors. The direct measurement methods are also unsuitable for determining the volume of inner defects. In contrast, the burn-off method can accurately measure Gc, the defect volume, and the non-uniformity of the external shape; hence, this method is recommended for ship design when the exact Gc and fabrication quality need to be known. There has been considerable research in other industries on fiber-reinforced composites. However, limited research has been conducted on the use of GFRPs in shipbuilding. Therefore, we expect that this comparative study on methods for assessing laminate structures of actual GFRP ship components will provide a good reference for GFRP ship design.

2. Theoretical Section

2.1. Mechanical Properties and Fabrication Defects of Laminate Structures

The bending strength of the laminate determines the thickness required to build a specific component. Figure 1 shows how the flexural strength of GFRP structures vary with Gc based on the results of estimation equations in ISO 12215-5 [6] and the rules of two classification societies—Lloyd's Register (LR) [9] and *Registro Italiano Navale* (RINA) [10]—in comparison to bending test results (according to ASTM D790) of actual GFRP laminates with 4- to 12-ply woven roving cloth with a weight per unit area of 570 g/m^2 [16]. The bending strength improves as Gc is increased. However, the experimental results are generally larger than the theoretical values owing to the safety margin included in the standards and classification rules. The test results indicate that the bending strength begins to decrease when Gc increases above 50 wt.% [6]. This is because a high Gc increases the probability of defects and resin-poor regions. The measured bending strengths of some samples with a Gc of 30–35 wt.% showed a significant decrease, which is likely because these specimens were taken from a heterogeneous section of laminate. In other words, to ensure the desired physical performance of laminate parts, the actual Gc should correspond to the design value. The fabrication quality is also important; Figure 2a shows the typical lamination defects that appear in fiber-reinforced composites, and Figure 2b shows inner defects in a carbon fiber-reinforced laminate detected by computed tomography (CT).

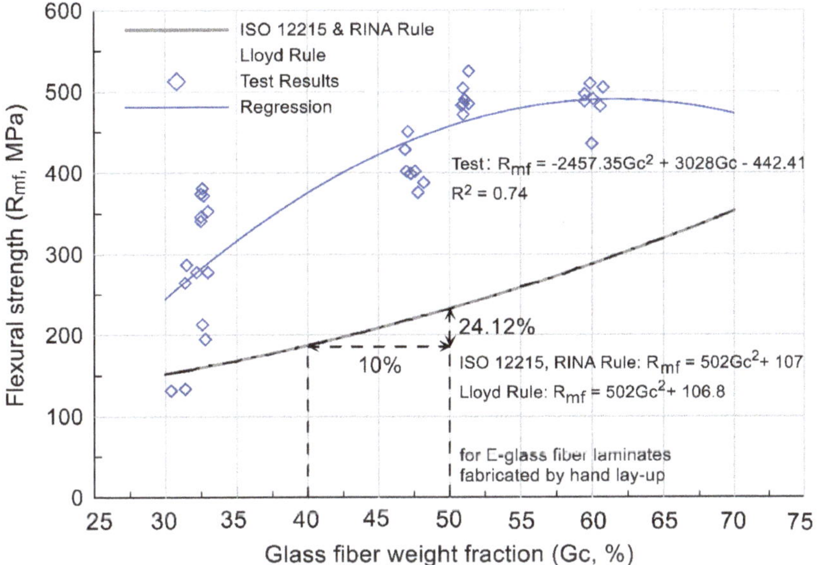

Figure 1. Variation in laminate flexural strength with Gc as determined by international rules and material tests.

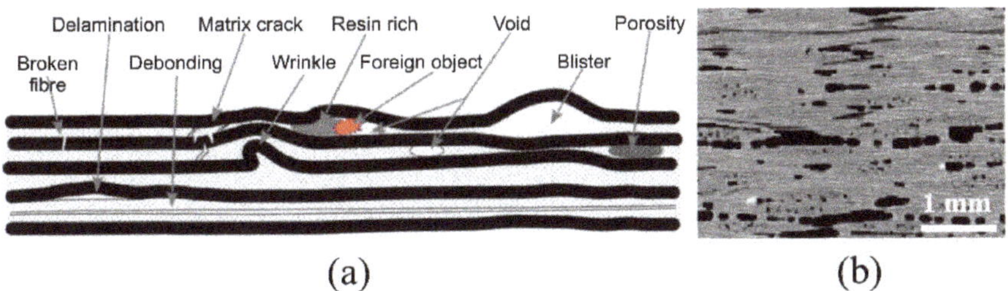

Figure 2. Defects in fiber-reinforced plastic laminates: (**a**) Schematic of common lamination defects [18]; (**b**) X-ray CT image of defects in carbon-fiber-reinforced plastic laminate [19].

2.2. Measurement of Weight Fraction of Glass Fiber (Gc) of Laminate Structure

As shown in Figure 2b, destructive methods such as CT testing allow the accurate measurement of inner defects such as void content. However, this method is not practical for the analysis of ship structures owing to the need for specialized equipment and technicians, which are typically not available during ship building. Therefore, we used the following four practical and simple methods to measure, compare, and analyze the GFRP laminate specimens:

1. Arithmetic calculation of Gc based on material design, as per ISO 12215 and other classification rules.
2. Calculation of the volume of laminate specimens using tools such as Vernier calipers, followed by calculation of Gc.
3. Measurement of the specific gravity of laminate specimens using Archimedes' principle, followed by calculation of Gc.
4. Burn-off method involving combustion of the matrix, as suggested by ASTM D3171-15, followed by weight measurement of the glass fiber to determine Gc.

This section summarizes the procedures for these four methods and their application to laminate structures.

2.2.1. Rule Calculation

International standards and classification societies provide rules for ship design, with formulas to calculate Gc and the mechanical properties of the laminate, as well as the required thickness of the laminate structure, by considering the Gc and the hull shape [7]. This process is called "scantling" in ship design. The rule calculations allow easy estimation of Gc from the properties of the fiber cloth and resin and are widely used for the structural design of composite ships. Gc can be calculated from the densities of the glass fiber and cured resin and the thickness of the laminate structure. ISO 12215 and classification society rules suggest that the glass fiber and resin densities should be taken as 2.56 and 1.2 g/cm^3, respectively, resulting in Equation (1) for calculating Gc [6].

$$\mathrm{Gc} = 2.56/(3.072 \times t/w + 1.36) \times 100, \quad (1)$$

where t is the laminate thickness (mm) and w is the weight of the glass fiber cloth per unit area (kg/m^2).

The thickness of a single impregnated ply, $T_{\mathrm{single\ ply}}$, can be obtained from the thickness of the laminate structure, t, as follows:

$$T_{\mathrm{single\ ply}} = w/3.072((2.56/\mathrm{Gc}) - 1.36), \quad (2)$$

where w is the weight of the glass fiber in a single ply of laminate (kg/m^2).

2.2.2. Direct Measurement

One of the simplest approaches for experimentally determining Gc is to measure the dimensions and weight of a section of the laminate structure using Vernier calipers and an electronic scale, respectively. By assuming a perfect cuboid shape (volume = cross − sectional area × thickness), the relative density, ϱ_c (g/cm^3), can be calculated (Equation (3)). The relative density can then be used to calculate Gc based on the theoretical densities of the glass fiber and resin (Equation (4)).

$$\varrho_c = M/(A \times t \times 1000), \quad (3)$$

$$\mathrm{Gc} = (p \times N \times 0.1)/(\varrho_c \times t) \times 100 \quad (4)$$

where M is the weight of the laminate specimen (g), A is the cross-sectional area (m^2), and t is the thickness (mm); p is the weight of one sheet of reinforcement per unit area (g/m^2), and N is the number of sheets in the specimen.

This method is both theoretically and experimentally simple. However, the volume and weight of the cut specimen must be measured accurately. The specimen density is often only approximated from measurements taken using Vernier calipers and a scale if the laminate specimen has rough or irregular surfaces; therefore, it is more accurate to measure the relative density directly using Archimedes' principle and an instrument such as an immersion electronic densimeter. ASTM D792-13 [20] proposes a procedure for measuring the relative density of plastics, which can be used as a reference. The measured result can then be used in Equation (4).

For both these methods, it is important that the measurements are accurate, and that the density of the glass fiber and uncured resin are known accurately. ISO 12215, LR, and RINA [9,10] prescribe the method described in ISO 1172 [21] for density measurements to confirm those reported in the catalog. However, the inner defects in the structure (see Figure 2) cannot be considered with either method. This can lead to an inaccurate measurement of Gc.

2.2.3. Matrix Burn-Off

To accurately measure Gc and the volume of inner defects, such as porosity, voids, and delamination, ISO 12215 and the classification society rules recommend the use of combustion methods whenever possible. However, this method is not widely used because the process is inconvenient and requires special equipment, and it is considered optional. ASTM D3171-15 [22] proposes two methods for determining the percentages of the raw materials contained in composite materials by removing the matrix. The first method is digestion, which involves chemical removal of the matrix phase, while the other involves the burn-off of the matrix in a furnace. The fibers used in GFRP ship structures can be deformed by the acid solvent used for digestion; therefore, the burn-off method is more appropriate.

After calcining the matrix of the laminate structure, Gc can be determined by comparing the weight of the specimen before and after combustion. This is calculated as follows:

$$\mathrm{Gc} = M_f/M_i \times 100, \tag{5}$$

where M_i and M_f are the weights (g) of the specimen before and after combustion, respectively.

The burn-off method has the advantage of allowing the matrix resin to be calcined slowly, which provides an accurate characterization of the quality of the laminate structure. Moreover, by dividing Equation (5) with the weight of the specimen before and after combustion, the volume of the voids can be obtained [22]:

$$V_{\mathrm{void}} = (1 - M_f/M_i \times \varrho_c/\varrho_f - (M_i - M_f)/M_i \times \varrho_c/\varrho_r) \times 100, \tag{6}$$

where ϱ_f and ϱ_r are the densities (g/cm^3) of the glass fiber and resin, respectively.

2.3. *Selection of Glass Fiber Content Measurement Method*

The principles and special properties of the methods described above for measuring the Gc of laminate structures for composite ships are summarized in Table 1. Among these methods, the digestion method was excluded because of the possibility of deformation of the raw material and because it is a time-consuming process [23]. To compare the other four methods in more detail, we applied each method to calculate Gc in laminate specimens taken from the side hull plate of a 16-m composite ship. The results are described and compared below.

Table 1. Comparison of determination methods for Gc.

Four Methods for Measuring Glass Fiber Weight Fraction			
Rule Calculation	Simple Direct Measurement	Method using Hydrometer	Burn-off
➢ ISO 12215 reports the densities of glass fiber and resin. Based on these values and the structure thickness, Gc can be calculated. ➢ This method is widely recommended by the classification societies. ➢ The calculation is simple, and the weight can be easily estimated from the thickness of the laminate. ➢ ISO 12215, ASTM D792	➢ Volume and weight are measured by using tools such as a Vernier caliper and a scale to calculate the relative density. ➢ Procedure is very simple. ➢ Volume errors of some external shapes can be estimated. ➢ ASTM D792	➢ The relative density of a cut laminate structure is measured by using a tool such as a densimeter. ➢ A relatively accurate Gc value can be obtained by using a simple tool. ➢ Interior defects such as voids cannot be considered. ➢ ASTM D792	➢ Weight of fiber in a laminate is measured by calcining a resin matrix using a furnace, and Gc is calculated based on the weight. ➢ Gc can be accurately calculated. ➢ A special experimental tool is required. ➢ Volume of fabrication defects in a laminate can be considered. ➢ ASTM D3171, ASTM D792

3. Experimental Methods

Fabrication of Composite Ship Hull Plate and Sample Characterization

The ship tested herein was a 16-m GFRP yacht with a displacement of 26 tons. The primary raw materials in the structure are chopped strand mat (CSM; weight of dry fabric per unit area: 450 g/m^2) and polyester resin. For the hull, which is the primary structure of the ship, 20-ply CSM was used for the bottom plate, and 16-ply CSM was used for the side plates. Figure 3 shows the primary structure of this ship, the structural layout of the hull, and the fabrication design for the hull plate.

For the composite structures used in the experiment, a hull side plate was fabricated with 16-ply CSM containing 40 wt.% glass fiber, in accordance with the fabrication design (Figure 4). Another prototype laminate structure based on the shape of the side plate, but with a different thickness, was designed by increasing the amount of glass fiber. The higher-Gc laminate structure was designed according to the ISO 12215-15 and RINA design rules; by intentionally increasing the design Gc to 64 wt.%, the number of plies of cloth was increased while the amount of resin was reduced to fabricate a composite hull structure with a higher Gc and smaller thickness. Both laminate structures were fabricated using the vacuum infusion method. Table 2 shows the material design of the two laminate structures, and Figure 4 shows the designs of the two prototypes and the fabricated hull plates. Four specimens were cut and prepared from each laminate structure.

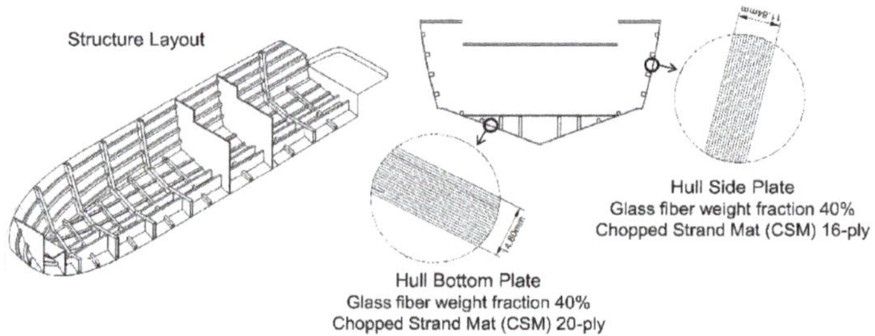

Figure 3. Details of the target ship structure and hull plate.

Figure 4. Dimensions of two hull plates and fabricated prototypes.

Table 2. Material design for fabrication of two hull side plates.

Item	Original Side Plate	High Gc Side Plate
Design Gc [wt.%]	40	64
Fiber type [density]	E-glass fiber [2.56 g/cm^3]	
Fabric type [weight per area]	Chopped strand mat (CSM) [450 g/m^2]	
Resin type [density]	Polyester [1.13 g/cm^3]	
Manufacturing thickness (mm)	11.84	9.36
No. of plies	16	24
Weight per area (kg/m^2)	18.00	16.88

4. Results

4.1. Specimen Size

The four specimens cut from each laminate structure are shown in Figure 5. Figure 6 shows detailed photos of the #1 laminate specimens with designed Gc values of 40 and 64 wt.%, respectively, indicating the defects generated during the fabrication process. The dimensions obtained for each specimen are listed in Tables 2 and 3.

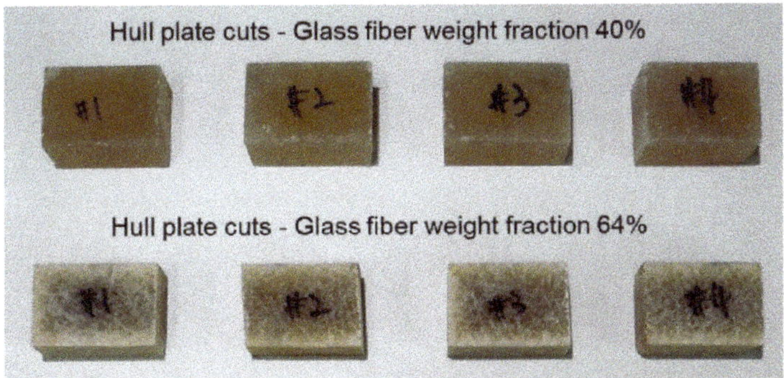

Figure 5. Specimens cut from two hull plates.

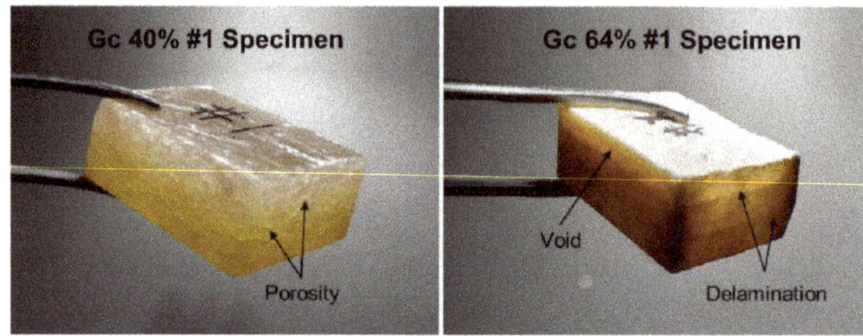

Figure 6. Detailed comparison of 40 and 64 wt.% Gc specimens.

Table 3. Dimensions of specimens taken from designed hull plate with Gc of 40 wt.%.

Sample	Length (mm)	Width (mm)	Thickness (mm)	Weight (g)
#1	27.46	20.19	11.70	9.67
#2	28.87	20.39	12.65	10.74
#3	26.38	20.22	12.05	9.66
#4	27.01	20.19	11.35	9.32
Mean			11.94	9.85

4.2. Gc Measurements

The four measurement methods summarized in Table 1 were used to determine Gc of the laminate sections with different design Gc values. The fabrication quality of the laminate structure was confirmed by establishing whether glass-fiber reinforcement was included as specified in the material design. The volume of defects and voids in the laminate structure was also determined. By comparing and analyzing the measurement results obtained using each method, we propose the most suitable method for use in optimized ship design.

4.2.1. Rule Calculation

When Gc is determined from the material design, it is possible to calculate the expected thickness and weight according to the method suggested by the rules; by working backward, Gc can be easily calculated from the fabricated structures. The thicknesses of the side plates of the hull were 11.84 mm for the Gc 40 wt.% design and 9.36 mm for the Gc 64 wt.% design. The side plates were scantled according to the RINA [10] rule (Table 2). The weight per square meter of the laminate structure can be calculated based on the material information used for the construction of this ship (Table 2). Furthermore, from Equation (2), the thickness of a single-ply with 40 wt.% Gc was determined to be 0.74 mm, whereas that with 64 wt.% Gc was 0.39 mm; when the ply number of each structure was applied, the weights of the two structures were 18 and 16.88 kg/m^2, respectively (Table 2). Using the calculations in Equation (1) yields Gc values for the two structures of 40 and 64 wt.%, respectively.

4.2.2. Simple Direct Measurement

For the simple direct measurement, the dimensions of the specimen were measured with Vernier calipers (Tables 3 and 4), and the weight was measured with an electronic scale (Figure 7). The results were used to calculate Gc of the specimens with Equations (3) and (4). In this process, some weights of the CSM cloth were also measured with an electronic scale and applied to the Gc calculation. Table 5 shows the calculated Gc results for the eight laminate specimens.

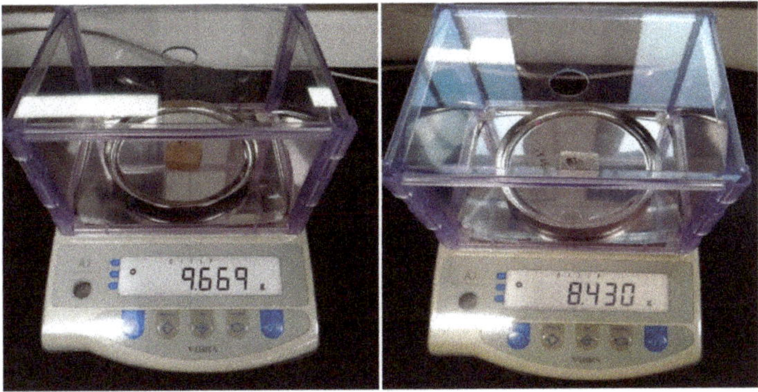

Figure 7. Weight measurement of specimens using an electronic scale.

Table 4. Dimensions of specimens taken from design hull plate with Gc of 64 wt.%.

Sample	Length (mm)	Width (mm)	Thickness (mm)	Weight (g)
#1	25.81	20.15	9.00	8.43
#2	29.18	20.01	9.07	9.56
#3	28.11	20.03	9.13	9.37
#4	27.15	19.90	9.22	8.59
Mean			9.11	8.99

Table 5. Gc calculation result for specimens based on simple direct measurements.

Sample	Glass Fiber Content (Gc) (wt.%)	
	Design Gc: 40 wt.%	Design Gc: 64 wt.%
#1	41.30	66.67
#2	39.53	65.79
#3	39.83	65.00
#4	42.01	68.10
Mean	40.67	66.39

Compared to the design Gc values of 40 and 64 wt.%, the simple direct measurements revealed average Gc differences of +0.67 and +2.39 wt.%, respectively. This variation is likely due to the differences between the densities of the glass fiber and polyester resin proposed in the rule and the actual values. In addition, errors occurring during laminate fabrication, perhaps resulting from the technical skills of a worker or a difference in shape volume, may also contribute to the discrepancy. The measured thicknesses of the laminate specimens (Tables 3 and 4) differed compared to the designed values by an average of +0.10 mm and −0.25 mm for Gc 40 and 64 wt.%, respectively. In other words, these results indicate that the error arises from differences in the external shape of the specimens, rather than from differences between the actual densities of the raw materials and the values specified in the rule.

4.2.3. Archimedes' Measurement

The relative densities of the hull plate specimens were measured in accordance with ASTM D792-13 [20] using a water immersion hydrometer with a precision of 0.01 g/cm^3 (Figure 8). For precise verification, the densities of the CSM cloth and cured polyester resin were also measured (Table 6); the average values were 2.56 and 1.22 g/cm^3, respectively, which are similar to the values suggested in the classification rules. The Gc of each hull plate specimen was calculated by using the measured density values (Tables 6 and 7) in Equation (2), and the results are listed in Table 8.

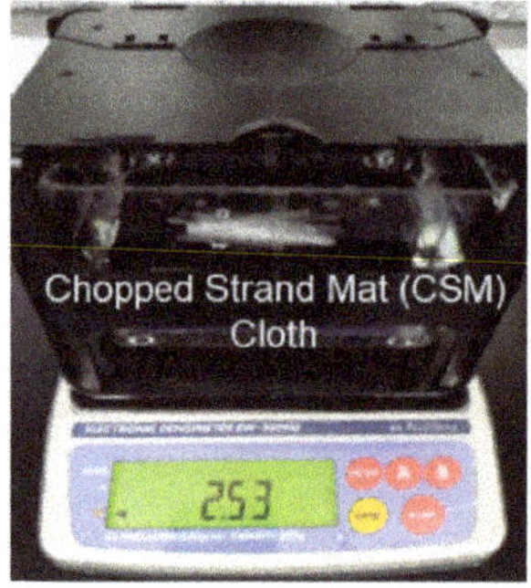

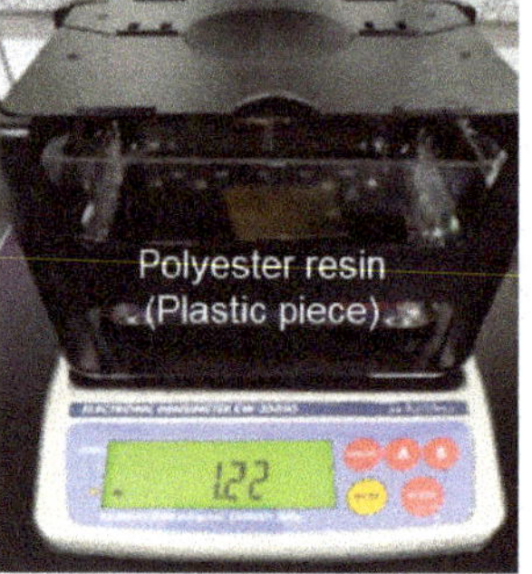

Figure 8. Measurement of the relative densities of CSM cloth, polyester, and hull plate using hydrometer.

Table 6. Densities of raw materials derived from hydrometer measurements.

Sample	Density (g/cm^3)	
	E-Glass Fiber	Polyester Resin (Cured)
#1	2.53	1.22
#2	2.57	1.22
#3	2.55	1.22
#4	2.60	1.22
Mean	2.56	1.22

Table 7. Measurement of specimen densities using a hydrometer.

Sample	Density (g/cm^3)	
	Design Gc: 40 wt.%	Design Gc: 64 wt.%
#1	1.53	1.78
#2	1.48	1.79
#3	1.52	1.78
#4	1.52	1.78
Mean	1.51	1.78

Table 8. Calculated Gc values of specimens measured using a hydrometer.

Sample	Glass Fiber Content (Gc) (wt.%)	
	Design Gc: 40 wt.%	Design Gc: 64 wt.%
#1	40.22	67.42
#2	38.46	66.52
#3	39.31	66.46
#4	41.73	65.81
Mean	39.93	66.55

For the specimens with design Gc of 40 and 64 wt.%, the measurement results showed differences of −0.07 and +2.55 wt.%, respectively, relative to the design Gc, and differences of −0.74 and −0.17 wt.%, respectively, relative to the average value of the Gc determined using the simple direct measurement. Overall, the results tend to be similar to those of the simple direct measurement; the slight difference is the result of volumetric error for the specimen. Therefore, the results measured using a hydrometer were more accurate. The volume measurement using the hydrometer indicates that there is no apparent defect, but it can be imagined that a higher vacuum was applied with a larger amount of E-glass fibers during the fabrication of the higher Gc laminate and that the thickness was slightly reduced. Overall, the direct measurement method indicates that the laminate structures were generally fabricated well and without significant quality defects resulting from incorrect scantling dimensions and non-uniformity of the external shape. In addition, the densities of the constituent materials are similar to those proposed in the design rules.

4.2.4. Burn-Off Method

In the burn-off method, a 4.5-L electric muffle furnace (CORE TECH, HQ-DMF 4.5, Figure 9) with a heating limit of 1200 °C was used to remove polyester from both laminate structures. The procedure described in ASTM D792-13 [20] was followed. The heating temperature and time were set to 600 °C and 3 h, respectively, in accordance with prior work utilizing muffle furnaces [12,24]. At 30-min intervals, the temperature inside the furnace and the weight of the specimen were measured, and then combustion was continued. This process was repeated until the weight change of the specimen was less than 0.001 g. Figures 9 and 10 summarize the burn-off test procedure for the laminate.

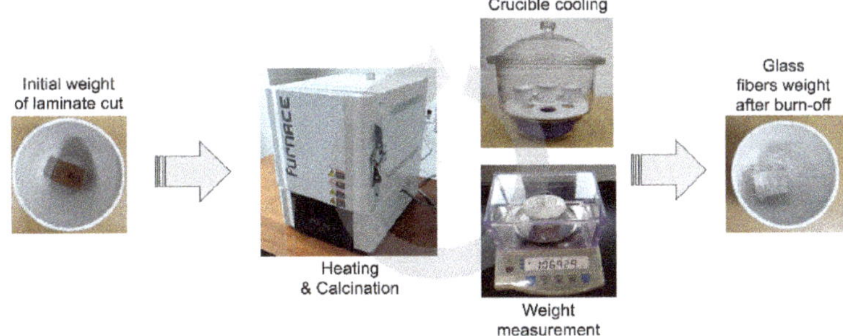

Figure 9. Burn-off method for the combustion of laminate specimens.

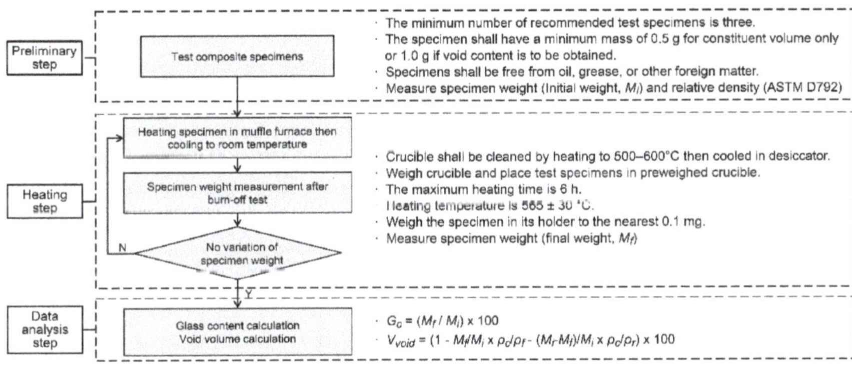

Figure 10. Burn-off test procedure for laminate structures, according to ASTM D792.

Figures 11 and 12 show the burn-off test results for specimen #1 with Gc 40 and Gc 64 wt.%. The resin was calcined rapidly during the first hour of heating. Furthermore, as the laminate was carbonized, the specimen gradually turned white. Tables 9 and 10 summarize the results of the burn-off tests and present the calculated Gc values based on the weight of the glass fiber remaining after the resin matrix was completely burned away.

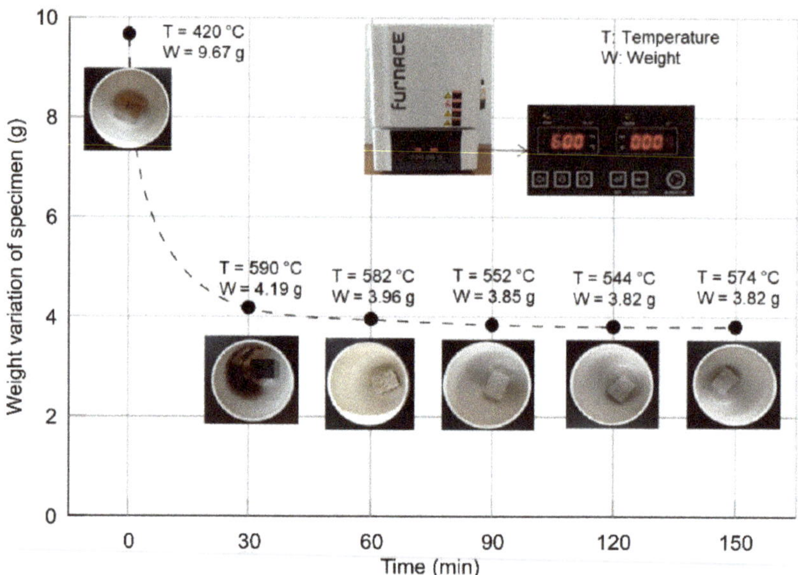

Figure 11. Changes in weight of specimen due to combustion (Gc 40 wt.%, Specimen #1).

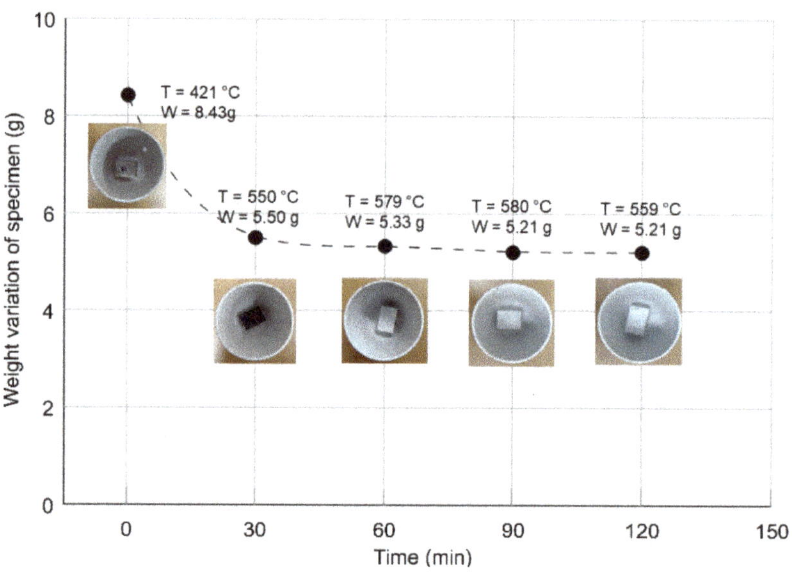

Figure 12. Changes in weight of specimens due to combustion (Gc 64 wt.%, Specimen #1).

Table 9. Gc calculation result obtained from the burn-off method (design Gc 40 wt.%).

Label	Specimen Weight (g)	Glass Fiber Weight (g)	Glass Fiber Content (wt.%)
#1	9.67	3.82	39.50
#2	10.74	3.83	35.66
#3	9.66	3.79	39.23
#4	9.32	3.65	39.16
Mean	-	-	38.39

Table 10. Gc calculation result obtained from the burn-off method (design Gc 64%).

Label	Specimen Weight (g)	Glass Fiber Weight (g)	Glass Fiber Content (wt.%)
#1	8.43	5.21	61.80
#2	9.56	5.93	62.03
#3	9.37	5.79	61.79
#4	8.59	5.31	61.82
Mean	-	-	61.86

The burn-off test confirmed that the specimens with a design Gc of 64 wt.% exhibited faster calcination of the resin, with complete burn-off completed approximately 30 min faster. Figure 13 shows the progression of the laminate weight change during combustion for each sample.

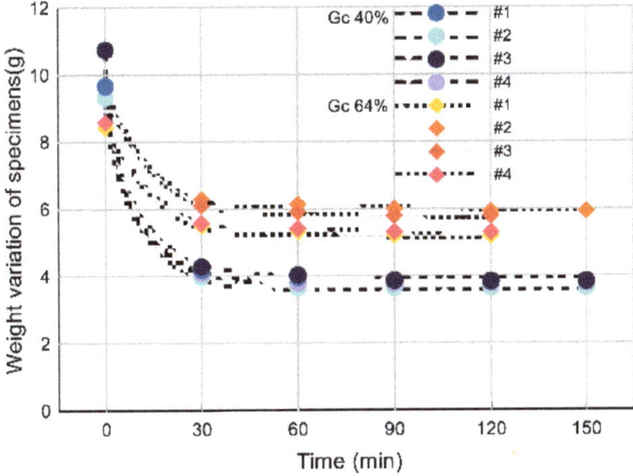

Figure 13. Changes in laminate weight during combustion.

The average Gc values measured using the burn-off method were 38.39 and 61.86 wt.% for the hull plates with design values of Gc 40 and 64 wt.%, respectively. These values respectively differ from the design values by −1.61 and −2.14 wt.%; from those obtained with the simple direct measurement by −2.28 and −4.53 wt.%; and from those obtained with the hydrometer method by −1.54 and −4.69 wt.%. Overall, it was confirmed that the weight fraction of the glass fiber was lower than the values obtained using the other measurement methods. This may be due to the volume of inner defects in the laminate (shown in Figure 2). The 40 wt.% Gc specimens seem to be affected by small porosities resulting from areas rich in resin, while in the 64 wt.% Gc specimens, the increase in the number of plies used may have caused delamination between the cloth layers and voids in resin-poor areas. As these specimens were fabricated by the infusion method and were well-fabricated without a significant quantity of defects, as was apparent from the previous hydrometer measurements, the overall void volume must be the total volume of many small voids. In particular, it was confirmed that an increased number of voids was present in the specimens of the hull plate with a high Gc.

Figure 14 shows the void volumes for the specimens of the two hull plates according to Equation (6). Although the volume fraction of the voids is small, this has a significant influence on the weight fraction of glass fiber.

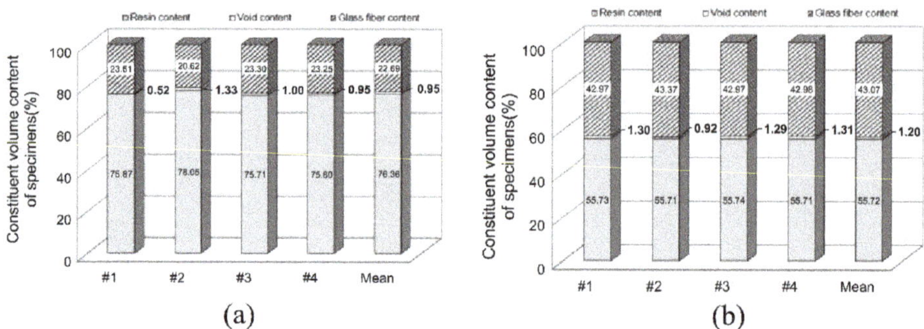

Figure 14. Volume fraction of voids in the specimens, measured by the burn-off method: (**a**) design Gc 40 wt.%; (**b**) design Gc 64 wt.%.

5. Discussion

The Gc values of the two hull plates with different design Gc were verified through four methods. Figure 15 shows a comparison of the experimental results. The differences between the Gc values obtained by the simple direct measurement and hydrometer method were not significant, because the error in the measurement of the external volume was relatively small. Therefore, the discrepancies in the results obtained using the burn-off method are likely due to differences in the volume of inner defects, such as voids. In other words, it was found that the volumes of hull plates with design Gc values of 40 and 64 wt.% contained voids with volumes comprising 0.95 and 1.20 wt.% of the total volume. As shown in Figure 14, the larger difference in the measured results for the Gc 64 wt.% hull plate can be considered to arise from the fabrication qualities of the laminate structures. Furthermore, although these are relatively well-fabricated prototypes produced using the vacuum infusion method, the differences in the results are not negligible. For example, the 2.14 wt.% difference in Gc obtained in the burn-off measurement method for the hull plate with a design Gc value of 64 wt.% corresponds to a decrease in the bending strength of the laminate of about 4.32%, based on the ISO 12215 equations.

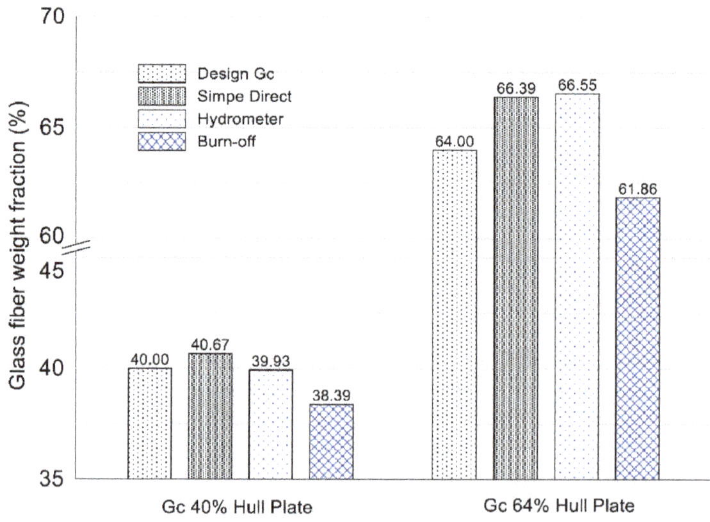

Figure 15. Comparison of Gc values measured by the rule calculation, direct measurement, and burn-off methods.

The experimental results are summarized in the form of a box plot, as shown in Figure 16. In the case of the 40 wt.% Gc hull plate, the shape, including the thickness, corresponded well with the design. Nevertheless, nonuniform impregnation of the polyester resin can be confirmed by the deviation in Gc value shown in Figure 15. In particular, as shown in Figure 14, specimen #2 (with a Gc of 35.66 wt.%) contained a relatively large volume of voids and polyester. In the case of the hull plate with a Gc of 64 wt.%, it can be confirmed from the increased Gc seen with the simple direct and hydrometer measurement methods that the pressure of the air compressor was slightly excessive during infusion. This was done to achieve the high Gc level, and it was confirmed that they were fabricated with a thickness slightly below the design thickness. However, this does not constitute an increase in the actual Gc value, but rather an increase in the internal void volume. This resulted in a lower E-glass fiber content, relative to the content specified in the design, and this was confirmed in the burn-off measurement results.

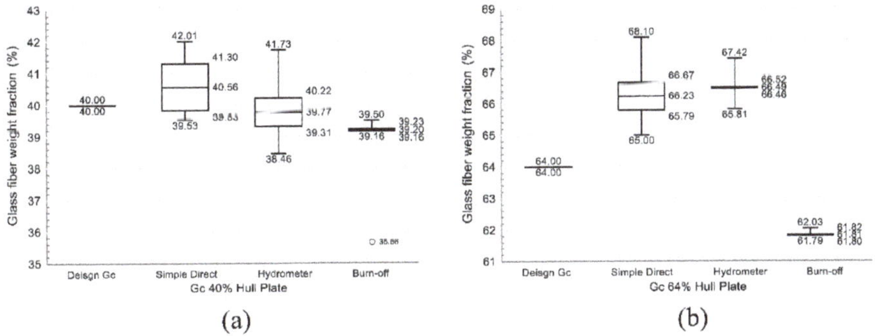

Figure 16. Gc distribution of specimens measured by different methods: (**a**) design Gc 40 wt.%; (**b**) design Gc 64 wt.%.

The results of each measurement method may be summarized as follows: the rule calculation can be used very quickly and efficiently if the ship design and material design are known. However, depending on the manufacturing quality of laminate structures, the errors in the measurements of physical properties may increase, and these errors cannot be confirmed. To reduce this error, it is reasonable to use the density suggested by the manufacturer or the density measured directly, rather than use the density of the fiber and resin suggested by the rule. In the simple direct measurement, the differences in shape between the design and laminate can be easily compared. In particular, it is easy to identify key errors from the comparisons of laminate thicknesses and weights. Nonetheless, it is preferable to use a device such as a hydrometer to identify apparent quality defects involving the density of raw materials or the non-uniformity of the external shape. However, with these methods, the effect of internal defects on Gc cannot be identified. Using the burn-off measurement method, suggested in the rules of classification societies, it was possible to measure Gc most accurately, and defects in laminates, including voids, could also be measured very accurately. However, this method requires specialized knowledge regarding analysis with the equipment used for combustion.

6. Conclusions

In this study, the Gc values of two hull plates of composite ships with different glass fiber weight fractions were measured using rule calculation, simple direct measurement, hydrometer, and burn-off methods, with the aim of identifying the change in fabrication quality resulting from quantitative changes in Gc. Because laminate structures for hull plates are typically fabricated by mixing two or more cloths in building yards, large differences can occur between the fabricated laminate structure and the designed structure. The following conclusions were obtained with these practical measurement tests.

Because the rule calculation suggested by classification societies is based on statistics and long-term experience, a logical and relatively accurate Gc value can be calculated for laminate structure design. For more accurate structure determinations, direct measurements using tools such as a hydrometer are recommended. However, only the combustion method was able to measure both the volume of internal defects and Gc with high accuracy.

Overall, it is reasonable to conclude that, in the case of a high-performance ship or a ship built using new materials, it is significantly important to ensure performance by verifying the weight fraction of glass fibers through the use of an accurate test method such as the burn-off method.

Author Contributions: Conceptualization, D.O.; Methodology, D.O.; Funding acquisition, D.O.; Manufacturing, J.N.; Test and investigation, Z.H. and S.J.; Writing—original draft, Z.H., and S.J.; Writing—review and editing, D.O. All authors have read and agreed to the published version of the manuscript.

Funding: This research was funded by the Basic Science Research Program through the National Research Foundation of Korea (NRF), funded by the Ministry of Education, grant number NRF-2017R1D1A3B03032051; and the IoT and AI based development of Digital Twin for Block Assembly Process Program of the Korean Ministry of Trade, Industry and Energy, Republic of Korea, grant number 20006978.

Conflicts of Interest: The authors declare no conflict of interest.

References

1. Shenoi, R.A.; Dulieu-Barton, J.M.; Quinn, S.; Blake, J.I.R.; Boyd, S.W. Composite materials for marine applications: Key challenges for the future. In *Composite Materials*; Nicolais, L., Meo, M., Milella, E., Eds.; Springer: London, UK, 2011; pp. 69–89. ISBN 978-0-85729-165-3.
2. Oh, D.K. Marine composites, FRP small craft, and eco-friendliness. *The Society of Naval Architects of Korea Webzine*. 2019. Available online: http://www.snak.or.kr/newsletter/webzine/news.html?Item=board21&mode=view&s_t=1&No=645 (accessed on 15 August 2019).
3. Mouritz, A.P.; Townsend, C.; Shah Khan, M.Z. Non-destructive detection of fatigue damage in thick composites by pulse-echo ultrasonics. *Compos. Sci. Technol.* **2000**, *60*, 23–32. [CrossRef]
4. Pickering, S.J. Recycling technologies for thermoset composite materials-current status. *Compos. Part A Appl. Sci. Manuf.* **2006**, *37*, 1206–1215. [CrossRef]
5. Conroy, A.; Halliwell, S.; Reynolds, T. Composite recycling in the construction industry. *Compos. Part A Appl. Sci. Manuf.* **2006**, *37*, 1216–1222. [CrossRef]
6. ISO (International Organization for Standardization). *ISO 12215-5: Small Craft—Hull Construction, and Scantlings—Part 5: Design Pressures for Monohulls, Design Stresses, Scantlings Determination*; ISO: Geneva, Switzerland, 2019.
7. Song, J.H.; Oh, D.K. Lightweight structure design for composite yacht with optimum fiber mass content. In Proceedings of the International SAMPE Symposium and Exhibition, Long Beach, CA, USA, 25 May 2016.
8. Jang, J.W.; Han, Z.Q.; Oh, D.K. Light-weight optimum design of laminate structures of a GFRP fishing vessel. *J. Ocean Eng. Technol.* **2019**, *33*, 495–503. [CrossRef]
9. LR (Lloyd's Register). *Rules & Regulations for the Classification of Special Service Crafts*; Lloyd's Register: London, UK, 2019.
10. RINA (Registro Italiano Navale). *Rules for the Classification of Pleasure Yachts. Part B—Hull and Stability*; Imago Media: Genova, Italy, 2019.
11. Oh, D.K.; Han, Z.Q.; Noh, J.K. Study on mechanical properties of CFRP multi-layered composite for ship structure in change with carbon fiber weight fraction. *Ship Ocean Eng.* **2019**, *48*, 85–88. [CrossRef]
12. Kedari, V.; Farah, B.; Hsiao, K.T. Effects of vacuum pressure, inlet pressure, and mold temperature on the void content, volume fraction of polyester/e-glass fiber composites manufactured with VARTM process. *J. Compos. Mater.* **2011**, *45*, 2727–2742. [CrossRef]
13. Ibrahim, M.E. Nondestructive testing and structural health monitoring of marine composite structures. In *Marine Applications of Advanced Fibre-Reinforced Composites*; Graham-Jones, J., Summerscales, J., Eds.; Woodhead Publishing: Cambridge, UK, 2016; pp. 147–183. ISBN 978-1-78242-250-1.
14. Abdelal, N. Effect of Voids on Delamination Behavior under Static and Fatigue Mode I and Mode II. Ph.D. Thesis, University of Dayton, Dayton, OH, USA, 2013.

15. Hakim, I.; Donaldson, S.L.; Meyendorf, N.; Browning, C.E. Porosity effects on interlaminar fracture behavior in carbon fiber-reinforced polymer composites. *Mater. Sci. Appl.* **2017**, *8*, 170–187. [CrossRef]
16. Han, Z.Q.; Jang, J.W.; Noh, J.K.; Oh, D.K. A Study on material properties of FRP laminates for a composite fishing vessel's Hull. In Proceedings of the KSPE Spring Conference, Manhattan, KS, USA, 6–8 June 2018; p. 113.
17. Kim, S.Y.; Shim, C.S.; Sturtevant, C.; Kim, D.; Song, H.C. Mechanical properties and production quality of hand-layup and vacuum infusion processed hybrid composite materials for GFRP marine structures. *Int. J. Naval Archit. Ocean Eng.* **2014**, *6*, 723–736. [CrossRef]
18. Bowkett, M.; Thanapalan, K. Comparative analysis of failure detection methods of composites materials' systems. *Sys. Sci. Cont. Eng.* **2017**, *5*, 168–177. [CrossRef]
19. Kastner, J.; Plank, B.; Salaberger, D.; Sekelja, J. Defect and Porosity Determination of Fibre Reinforced Polymers by X-ray Computed Tomography. In Proceedings of the 2nd International Symposium on NDT in Aerospace 2010-We.1.A.2, Hamburg, Germany, 22–24 November 2010.
20. ASTM D792-13. *Standard Test Methods for Density and Specific Gravity (Relative Density) of Plastics by Displacement*; ASTM International: West Conshohocken, PA, USA, 2013. [CrossRef]
21. ISO (International Organization for Standardization), ISO 1172. *Textile-Glass-Reinforced Plastics-Prepregs, Moulding Compounds, and Laminates-Determination of the Textile-Glass and Mineral-Filler Content-Calcination Methods*; ISO: Geneva, Switzerland, 1996.
22. ASTM D3171-15. *Test Methods for Constituent Content of Composite Materials*; ASTM International: West Conshohocken, PA, USA, 2015. [CrossRef]
23. He, H.W.; Huang, W.; Gao, F. Comparison of four methods for determining fiber content of carbon fiber/epoxy composites. *Int. J. Polym. Anal. Charact.* **2016**, *21*, 251–258. [CrossRef]
24. McDonough, W.; Dunkers, J.; Flynn, K.; Hunston, D. A test method to determine the fiber and void contents of carbon/glass hybrid composites. *J. ASTM Int.* **2004**, *1*, 1–15. [CrossRef]

Article

An Analytical Model for the Tension-Shear Coupling of Woven Fabrics with Different Weave Patterns under Large Shear Deformation

Yanchao Wang [1,2,3,*], Weizhao Zhang [4], Huaqing Ren [5], Zhengming Huang [2], Furong Geng [1], Yongxiang Li [1] and Zengyu Zhu [6]

1. Research & Design Center, Guangzhou Automobile Group Co. Ltd., Guangzhou 511434, China; gengfurong@gacrnd.com (F.G.); liyongxiang@gacrnd.com (Y.L.)
2. School of Aerospace Engineering & Applied Mechanics, Tongji University, Shanghai 200092, China; huangzm@tongji.edu.cn
3. School of Mechanical and Automotive Engineering, South China University of Technology, Guangzhou 510006, China
4. Department of Mechanical and Automation Engineering, Chinese University of Hong Kong, Hong Kong, China; weizhaozhang2014@u.northwestern.edu
5. Department of Mechanical Engineering, Northwestern University, Evanston, IL 60208, USA; huaqingren2013@u.northwestern.edu
6. Guangdong Yatai New Material Technology Co. Ltd, Sihui City 526241, China; ceo@ytbxg.com

* Correspondence: wangyanchao@gacrnd.com

Received: 13 December 2019; Accepted: 25 January 2020; Published: 24 February 2020

Featured Application: The work provides an analytical model which can be useful in the optimization of process parameters of dry fabrics in material-forming industries.

Abstract: It is essential to accurately describe the large shear behavior of woven fabrics in the composite preforming process. An analytical model is proposed to describe the shear behavior of fabrics with different weave patterns, in which tension-shear coupling is considered. The coupling is involved in two parts, the friction between overlapped yarns and the in-plane transverse compression between two parallel yarns. By introducing the concept of inflection points of a yarn, the model is applicable for fabrics with different weave patterns. The analytical model is validated by biaxial tension-shear experiments. A parametric study is conducted to investigate the effects of external load, yarn geometry, and weave structure on the large shear behavior of fabrics. The developed model can reveal the physical mechanism of tension-shear coupling of woven fabrics. Moreover, the model has a high computational efficiency due to its explicit expressions, thus benefiting the material design process.

Keywords: analytical model; fabrics; weave pattern; shear deformation; tension-shear coupling

1. Introduction

Woven fabrics and their composites have been widely used in engineering due to the outstanding mechanical performance [1,2]. Recently, wet compression molding of woven composites has attracted more attention because of its high efficiency in the manufacturing process [3]. The shear deformation, the dominant mode of a woven fabric in the forming process, is closely related to the yarn reorientation which is critical for the mechanical properties of a composite. It is noted that a blank holder is commonly used to provide a tensile force in a preforming process so that wrinkle onset can be postponed [4]. Moreover, the shear stiffness of a woven fabric can be significantly affected by applying a tensile force along yarns [5,6]. Thus, it is critical to establish a model that can account for tension-shear coupling for the large shear deformation of a woven fabric.

Approaches to investigate the shear behavior of a woven fabric can be classified into three categories [7]: experimental testing, numerical simulation, and analytical modeling. In the first category, many kinds of experimental configurations have been reported [8,9]. For the second category, numerical simulations of woven fabrics have also been widely investigated, for example, Boisse et al. [10], Zhang et al. [11], and Erol et al. [12]. Experiments are essential for the understanding of mechanical properties of materials. However, it is too expensive to obtain all the properties of a material under various load cases only upon experiments. Conversely, numerical simulations and analytical models can predict mechanical behaviors of a material based on some basic test results, which significantly accelerates the application process of a material. Furthermore, compared with numerical methods, analytical models are superior in computational efficiency and understanding of physical mechanism. Thus, this work mainly focuses on analytical models.

Analytical models for shear behavior of woven fabrics can be split into two categories: kinematic models and mechanical models. The kinematic models [13,14] are succinct, but they cannot account for the effect of mechanical behaviors of yarns on the deformation of woven fabrics. To tackle this problem, mechanical models were developed. Mechanical models can be roughly divided into two families: continuum models and discrete models. With respect to the former family, King et al. [15] proposed a continuum mechanics model for woven fabrics, where the bending, tension, and shear behaviors of yarns are modeled by virtual springs. The parameters of the virtual springs are obtained from experiments directly, meaning that the coupling between tension and shear properties is only involved implicitly. Peng et al. [16] developed a continuum material model for woven fabrics, where the shear and tension stresses are decoupled by using a non-orthogonal coordinate system. However, the interaction between shear and tension properties still exists but is not considered. Gong et al. [17] proposed an anisotropic hyper-elastic model for the preforming process of woven fabrics, where the tension-shear coupling effect was studied. Nevertheless, the coupling factors are directly obtained by fitting experimental data, meaning that the coupling mechanism is not fully revealed.

In the range of discrete models, Lomov and Verpoest [18] presented an analytical model to investigate the tension-shear coupling in woven fabrics. In their work, the effect of transverse in-plane compression of yarns on the shear property is considered indirectly by the friction dissipation energy, which may underestimate its contribution. Zhu et al. [19] pointed out that the shear dissipation energy can be divided into three parts: yarn sliding, yarn compression, and yarn rotation. However, no tensile forces along yarns were studied in the work of Zhu et al. [19]. Erol et al. [20] gave analytical expressions for the tensile, compressive, and shear properties of woven fabrics by treating the structure as a system of beams and springs. Similarly, Nasri et al. [21] developed a hybrid discrete hypo-elastic model for woven fabrics, where the tensile and shear properties are modeled by tensile and torsional springs, respectively. The simulation results of both Erol et al. [20] and Nasri et al. [21] agree well with experimental data. However, in their work, the constitutive equations for the mechanical properties of woven materials are given by curve-fitted exponential or polynomial functions. Such operations are inconvenient when the weave pattern changes with design requirements.

To the authors' knowledge, the tension-shear coupling related parameters in most models must be determined by fitting experimental data, thereby needing to redo experiments when the weave pattern changes. In this work, an analytical approach is proposed to model the large shear deformation of woven fabrics. The tension-shear coupling mechanism is explicitly revealed by the friction dissipation energy and transverse compression dissipation energy. In addition, the effect of the weave pattern on the shear behavior of a fabric is considered by introducing the number of yarn inflection points. Thus, the model is applicable for fabrics with different weave patterns without more experiments required.

2. Tension-Shear Coupling Model

2.1. Geometrical Model

Figure 1 shows the configuration of a plain-woven representative volume element (RVE). The RVE geometric parameters, w_0, g_0, and N_y, are, respectively, the yarn width, the yarn gap, and the yarn amount in the RVE. L_R, the side length of the RVE, is defined as follows:

$$L_R = N_y(w_0 + g_0). \tag{1}$$

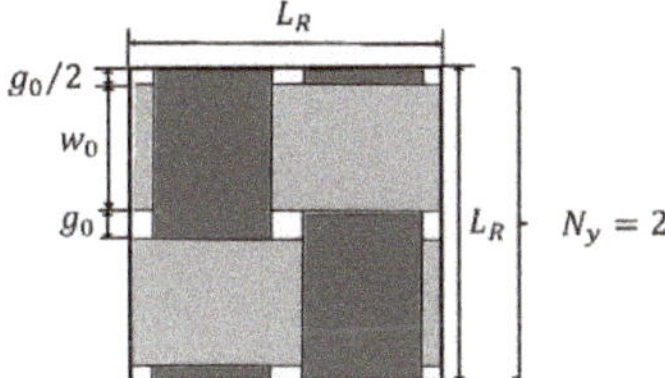

Figure 1. Geometry of a plain-woven RVE.

In order to focus on the tension-shear coupling, this work only considers the case that the warp and fill yarns have identical geometry. It should be noted that Equation (1) is applicable for fabrics with different weave patterns. The plain weave pattern shown in Figure 1 is only an example for illustration.

The number of inflection points of a yarn, η_y is introduced to characterize different weave patterns. In a woven fabric, a warp yarn passes over and under fill yarns in sequence, and vice versa. The ratio of the under and over areas of warp and fill yarns determines the weave pattern. In this work, when a yarn turns from under to over or from over to under one time, it is counted as one inflection point. The inflection point at the boundary of an RVE is counted as a half point. For example, as shown in Figure 2a, the inflection point number of a warp yarn is 2. The weave pattern of RVEs with the same yarn amounts can be distinguished by different yarn inflection point numbers. For example, as shown in Figure 2a,b, both the twill and plain RVEs have 4 warp and 4 fill yarns, but their inflection point numbers of the warp yarn are 2 and 4, respectively. The number of inflection points in an RVE determines the normal pressure between overlapped yarns induced by the tensile force. Furthermore, normal pressure is closely related to the tension-shear coupling which will be explained in detail hereinafter. Hence, it is believed that the inflection point number is critical to evaluate the effects of different weave patterns on the tension-shear coupling of fabrics.

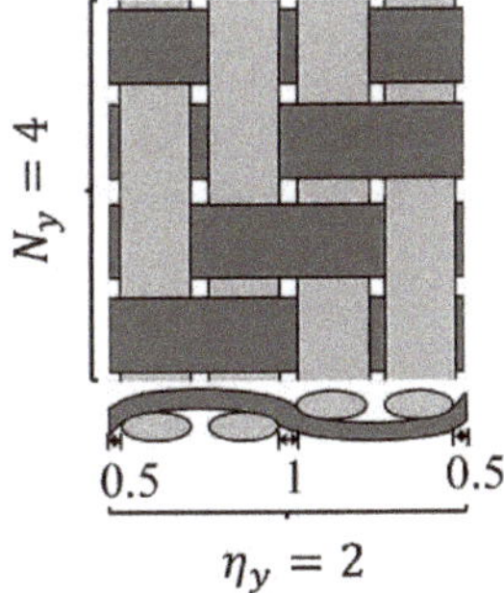

(**a**) Inflection points of a twill-woven RVE

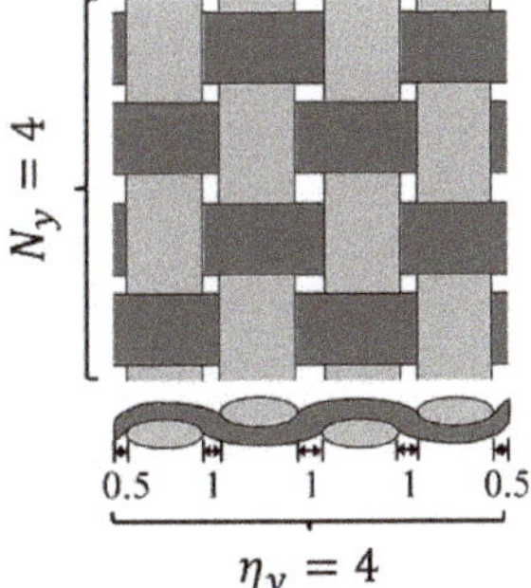

(**b**) Inflection points of a plain-woven RVE

Figure 2. Illustration of inflection points.

2.2. Incremental Energy Equilibrium Theory

In forming manufacturing industries, the shape and stress state of a formed fabric depend on the forming process parameters, such as molding shape, forming rate, etc. Numerical simulations are generally used to optimize such forming process parameters. The key to such a simulation is a reliable constitutive relation that describes the mechanical behavior of an RVE of a fabric. In practice, blank-holders are often used to introduce tension forces along yarns so that the wrinkling phenomenon may be postponed. Besides, a fabric is majorly shear deformed in the forming process. Thus, it is critical to establish a constitutive relation for large shear deformation of a fabric RVE with the consideration of tension-shear coupling. Once the constitutive relationship is obtained, the numerical simulation can be realized subsequently. Thus, this work mainly investigates a fabric RVE subjected to a combined tension and shear load, as shown in Figure 3.

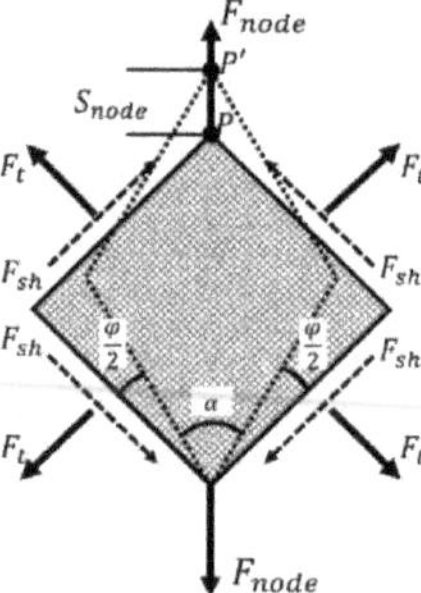

Figure 3. Shear deformation of a fabric RVE.

The RVE is applicable for different weave patterns. Thus, no specific weave pattern is illustrated in Figure 3. The RVE is firstly subjected to tension loads along yarns, F_t. The tension forces remain constant throughout the load case. Then, the RVE undergoes shear deformation under the force F_{node} at the node P of the RVE. The node P moves to P′ due to the shear deformation. The shear force F_{sh} is the projection of F_{node} to the boundary of the RVE. The relationship between F_{sh} and F_{node} is given as follows:

$$F_{sh} = \frac{F_{node}}{2\cos\left(\frac{\pi}{4} - \frac{\varphi}{2}\right)}. \quad (2)$$

As mentioned in the above paragraph, the tension forces along yarns are applied before the shear deformation, meaning that a yarn has been stretched before the RVE is sheared. Thus, no tension input energy is involved in the shear deformation. The input energy generated by the node force F_{node} will be dissipated by friction, transverse compression, thickness variation, yarn elongation, yarn sliding, and waviness angle variation. Among these dissipation energies, the former two are dominant. The yarn thickness may increase due to in-plane transverse compression, but the variation is small since the increase of yarn thickness will be restricted by overlapped yarns. The yarn elongation is negligible, since the longitudinal modulus of a yarn is much larger than the shear and transverse compressive properties of a fabric. The yarn sliding is also negligible, since a couple of balanced tension forces are applied to the end of a yarn. The waviness angle variation mainly results from yarn sliding and thickness variation, thereby also being insignificant. Thus, only the friction and transverse compression dissipation energies are considered. According to the energy equilibrium theory, the input energy should equal the sum of the dissipation energies. Hence, the following equation is obtained:

$$\mathrm{d}W_{node} = \mathrm{d}W_{fri} + \mathrm{d}W_{com}, \quad (3)$$

where dW_{node} is the input energy increment generated by the node force, F_{node}. dW_{fri} and dW_{com} are incremental energy dissipated by the respective friction and transverse compression of yarns.

Our goal is to express all items in Equation (3) as functions of shear force, shear angle, geometric parameters, and mechanical properties of yarns so that a shear stress-angle curve can be obtained eventually.

2.3. Input Energy of Shear Force

As shown in Figure 3, when a fabric RVE undergoes shear deformation, the node P moves to P' with displacement of S_{node}. The yarn angle α decreases with the increase of shear angle φ, where a relation exists: $\alpha = \pi/2 - \varphi$. Then, the incremental input energy is given by

$$dW_{node} = F_{node} dS_{node}. \tag{4}$$

In order to characterize the shear behavior of the RVE, it is necessary to express F_{node} and dS_{node} as functions of the shear stress and shear angle. Firstly, with Equation (2), the following equation is given as:

$$F_{node} = 2cos\left(\frac{\pi}{4} - \frac{\varphi}{2}\right)F_{sh}. \tag{5}$$

Furthermore, for the purpose of normalization, the shear stress τ_{sh} instead of the shear force F_{sh} is employed. Then, we have

$$F_{sh} = 2\tau_{sh} L_R t, \tag{6}$$

where L_R is the side length of the RVE, and t is the yarn thickness. Secondly, dS_{node} is the differentiation of the node displacement S_{node}, whose relationship is given by

$$dS_{node} = \frac{d(S_{node})}{d\varphi} d\varphi. \tag{7}$$

From Figure 3, S_{node} is given by

$$S_{node} = 2L_R \cos\left(\frac{\pi}{4} - \frac{\varphi}{2}\right) - 2L_R cos\frac{\pi}{4}. \tag{8}$$

Then, dS_{node} is given by

$$dS_{node} = \frac{d(S_{node})}{d\varphi} d\varphi = L_R sin\left(\frac{\pi}{4} - \frac{\varphi}{2}\right)d\varphi. \tag{9}$$

Finally, substituting Equations (5), (6), and (9) into Equation (4), the input energy is expressed in terms of the shear stress and the geometric parameters as follows:

$$dW_{node} = 4tL_R^2 \tau_{sh} \cos\left(\frac{\pi}{4} - \frac{\varphi}{2}\right) sin\left(\frac{\pi}{4} - \frac{\varphi}{2}\right)d\varphi = 2tL_R^2 \tau_{sh} sin\left(\frac{\pi}{2} - \varphi\right)d\varphi = 2tL_R^2 \tau_{sh} cos\varphi d\varphi. \tag{10}$$

2.4. Dissipation Energy of Friction

Figure 4a is a crossover point of two yarns. As shown in Figure 4b, a friction force is generated when a yarn rotates. An equivalent circle is used to approximate the contact area of overlapped yarns. It is reasonable to assume that the friction force is evenly distributed in the contact area. Imagine an infinitesimal sector whose side lengths are dr and $rd\phi$, as shown in Figure 4c. The sector moves from A to A' with the yarn rotation. Then, the moment of the friction force to the point O is given by

$$M_{fri} = \int_0^{2\pi} \int_0^R \frac{F_{fri}}{\pi R^2} r \cdot r dr d\phi = \frac{2}{3} F_{fri} R. \tag{11}$$

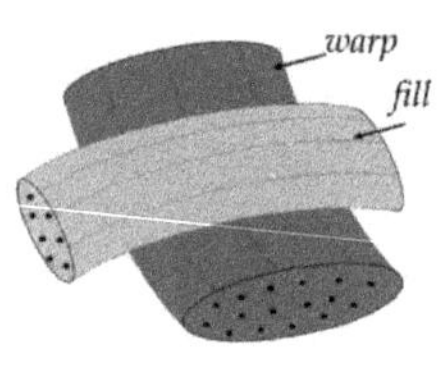

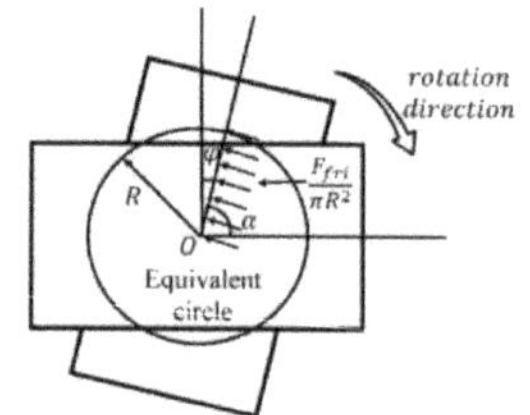

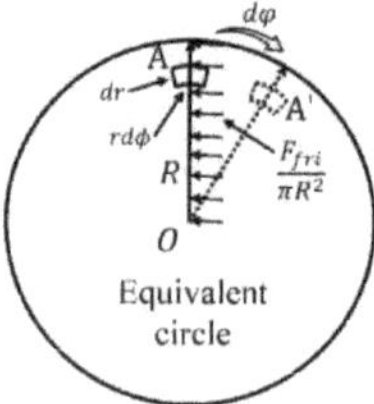

(**a**) Cross point in an RVE (**b**) Illustration of friction force (**c**) Calculation of friction energy

Figure 4. Dissipation energy of friction.

Furthermore, the friction dissipation energy in an RVE is obtained as

$$dW_{fri} = N_y^2 M_{fri} d\varphi = \frac{2}{3} N_y^2 F_{fri} R d\varphi, \tag{12}$$

where R is the equivalent radius. The contact area varies with the shear angle until parallel yarns contact each other in in-plane transverse direction. Thus, R is expressed as

$$R = \begin{cases} \frac{w_0}{\sqrt{\pi \cos\varphi}} & \varphi < \varphi_L \\ \frac{w_0}{\sqrt{\pi \cos\varphi_L}} & \varphi \geq \varphi_L \end{cases}, \tag{13}$$

where φ_L is the shear-locking angle, meaning that parallel yarns start to contact each other in in-plane transverse direction. The shear-locking angle can be expressed as

$$\cos\varphi_L = \frac{w_0}{w_0 + g_0}. \tag{14}$$

It should be noted that the shear angle keeps increasing after φ_L due to in-plane transverse compression of yarns. In other words, the shear angle is actually not locked. However, following the nomenclature of textile forming, the shear-locking angle is still employed.

For simplicity, Coulomb's friction theory is employed. Before a yarn rotates, the static friction force can be approximately seen as linearly dependent on the shear angle. When a yarn rotates, the friction force F_{fri} is calculated from the friction factor μ and the contact force between overlapped yarns F_c. The friction force is expressed as:

$$F_{fri} = \begin{cases} \frac{\varphi}{\varphi_0} \mu F_c & \varphi < \varphi_0 \\ \mu F_c & \varphi \geq \varphi_0 \end{cases}, \tag{15}$$

where φ_0 is the critical shear angle when a yarn starts to rotate. Obviously, the friction force depends on the contact force between overlapped yarns, F_c. F_c consists of two parts: the one induced by the tension force along a yarn, F_t^c, and the initial contact force induced by the yarn bending, F_0^c, as shown in the following equation:

$$F_c = F_t^c + F_0^c. \tag{16}$$

Ideally, F_0^c should be calculated from the yarn bending stiffness and the bending angle. However, the measured bending stiffness has large scatter, and its value is sensitive to the load status of a yarn. Thus, it is better to curve-fit F_0^c by comparing prediction results to experimental data.

Figure 5 illustrates the contact force induced by the tension force along a yarn for a plain-woven fabric. θ is the waviness angle of a yarn. L_{in} is the length of the inflected part of a yarn. Then, F_t^c is:

$$F_t^c = 2F_t sin\theta = 2F_t \frac{t}{L_{in}} = \frac{2tF_t}{\sqrt{(w_0 + g_0)^2 + t^2}}. \tag{17}$$

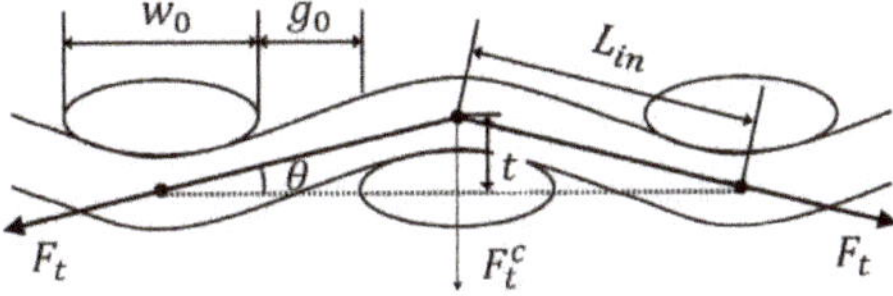

Figure 5. The contact force produced by the tension force of a yarn (plain-woven fabric).

For fabrics with different weave patterns, the illustration of the contact force is shown in Figure 6.

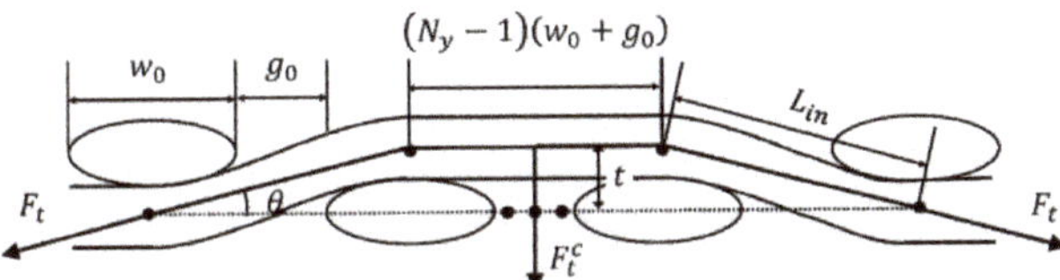

Figure 6. The contact force produced by the tension force of a yarn (different weave pattern).

Then, the contact force at one cross point is:

$$F_t^c = \frac{2\eta F_t sin\theta}{N_y} = \frac{2\eta_y t F_t}{N_y\sqrt{(w_0 + g_0)^2 + t^2}}. \tag{18}$$

The friction force can be calculated by substituting Equation (18) into Equation (16) and then into Equation (15).

Finally, the dissipation energy, dW_{fri}, is obtained by substituting Equations (15) and (13) into Equation (12). The expression of dW_{fri} is obtained:

$$dW_{fri} = \begin{cases} \frac{2}{3}N_y^2\frac{\varphi}{\varphi_0}\mu\left(\frac{2\eta_y t F_t}{N_y\sqrt{(w_0+g_0)^2+t^2}} + F_0^c\right)Rd\varphi & \varphi < \varphi_0 \\ \frac{2}{3}N_y^2\mu\left(\frac{2\eta_y t F_t}{N_y\sqrt{(w_0+g_0)^2+t^2}} + F_0^c\right)Rd\varphi & \varphi \geq \varphi_0 \end{cases}. \tag{19}$$

2.5. Dissipation Energy of in-plane Transverse Compression

As shown in Figure 7a, parallel yarns start to contact each other at the shear-locking angle. As the shear angle increases, the yarns are transversely compressed under the in-plane transverse compression force F_{com}. The points C and D move to C′ and D′, and the total width reduction of a yarn is Δw, as illustrated in Figure 7b. Since no yarn sliding is considered in this work, the side length of the parallelogram ABCD remains constant, which is $(w_0 + g_0)$.

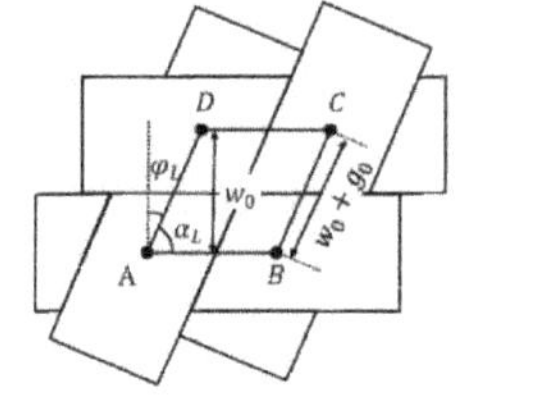

(a) Geometry at the shear-locking angle

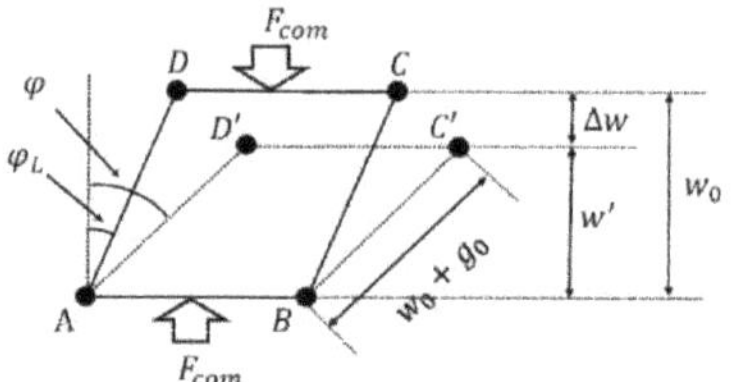

(b) Geometry after the shear-locking angle

Figure 7. In-plane transverse compression of an RVE.

In an infinitesimal load increment, F_{com} can be seen as constant. Since the fill and warp yarns are consistent in geometry and mechanical properties, the dissipation energy of the in-plane transverse compression of a fabric RVE is

$$\mathrm{d}W_{com} = 2N_y^2 F_{com} dw, \tag{20}$$

where dw is the differentiation of Δw. Δw and dw are given by

$$\Delta w = w_0 \varepsilon_{com}, \tag{21}$$

$$dw = w_0 d\varepsilon_{com}, \tag{22}$$

where ε_{com} is the in-plane transverse compressive strain of a yarn. From the view point of simplicity, it is reasonable to assume that the yarn compression is linear. Then, the compression force F_{com} is calculated with the following equation:

$$F_{com} = t(w_0 + g_0) E_{com} \varepsilon_{com}, \tag{23}$$

where E_{com} is the transverse compressive modulus of a yarn.

Please remember that our goal is to obtain the constitutive relation between the shear stress and shear angle of a fabric RVE. Thus, ε_{com} must be written as a function of the shear angle. φ_L is the shear-locking angle given by Equation (14). Obviously, when $\varphi \leq \varphi_L$, the width is not compressed, and both ε_{com} and $d\varepsilon_{com}$ are zero. When $\varphi > \varphi_L$, is the following equation is given as:

$$\Delta w = w_0 - w' = (w_0 + g_0)(cos\varphi_L - cos\varphi), \quad \varphi > \varphi_L. \tag{24}$$

Then, ε_{com} is obtained as

$$\varepsilon_{com} = \frac{\Delta w}{w_0} = \frac{w_0 - w'}{w_0} = \frac{(w_0 + g_0)(cos\varphi_L - cos\varphi)}{w_0}, \ \varphi > \varphi_L. \tag{25}$$

Furthermore, $d\varepsilon_{com}$ is reached:

$$d\varepsilon_{com} = \frac{d(\varepsilon_{com})}{d\varphi}\mathrm{d}\varphi = \frac{(w_0 + g_0)sin\varphi}{w_0}d\varphi, \varphi > \varphi_L. \tag{26}$$

Obviously, when $\varphi \leq \varphi_L$, the dissipation energy is zero. When $\varphi > \varphi_L$, $\mathrm{d}W_{com}$ can be achieved by substituting Equations (25) and (26) into Equations (22) and (23) and then into Equation (20). Finally, the expression of $\mathrm{d}W_{com}$ is achieved:

$$\mathrm{d}W_{com} = 2N_y^2 E_{com} \frac{t(w_0 + g_0)^3}{w_0}(cos\varphi_L - cos\varphi)sin\varphi d\varphi, \quad \varphi > \varphi_L. \tag{27}$$

2.6. Model Summary

Both the input energy and dissipation energies of friction and transverse compression are expressed as functions of the shear stress and shear angle. Then, the relation between shear stress and shear angle can be obtained by solving the energy equilibrium, Equation (3). It is noted that the friction dissipation energy is sensitive to the number of inflection points in an RVE, indicating that the present model can consider the effect of different weave patterns on the shear deformation of fabrics. In addition, the tension-shear coupling mechanism appears from two aspects in the present model. Firstly, the effect of tension force on the shear property of a fabric is directly reflected by the shear friction dissipation energy. An increase in the tension force leads to the increase of contact force between overlapped yarns and then results in the increase of friction dissipation energy. Obviously, a larger dissipation energy means larger resistance in deformation. Secondly, the tension force will make a yarn shrink. Generally, the transverse compressive modulus will rise due to the shrinking effect. Furthermore, the increase of transverse compressive modulus also generates more dissipation energy. With the help of the analytical model developed in this work, the effect of tension on the shear behavior of fabrics can be analyzed explicitly.

3. Model Validation

In this part, the present model is validated by experimental results of a dry woven fabric under tension and shear loads. The experimental data for a carbon fiber plain-woven fabric subjected to biaxial tension and shear force were presented in Nosrat-Nezami et al. [22]. As shown in Figure 8, the experiment is constructed by adding a tensile device to a picture frame configuration, and a needle bar is set up between clamps to guarantee the free rotation of yarns. The geometric information of the test specimen is shown in Table 1. The friction factor between warp and fill yarns of a carbon fabric is 0.6 according to Najjar et al. [23] and Allaoui et al. [24]. Please note that the width and gap of warp and fill yarns are slightly different. The shear-locking angle is determined by geometry of the yarn family which is first in-plane transversely contacted. Thus, the width and gap of the fill yarn are used in the present model.

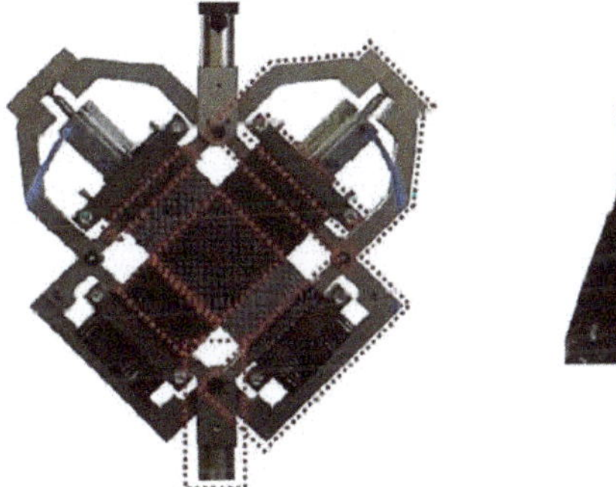
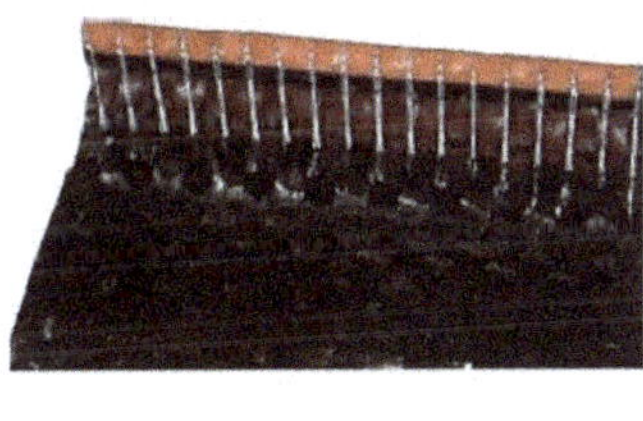

Figure 8. Biaxial tension and shear test on a plain-woven carbon fabric [22].

Table 1. Geometric information of the T700-12K carbon fiber plain-woven fabric specimen [22].

Specimen Size	Fabric Thickness	Warp Width	Warp Gap	Fill Width	Fill Gap
100 × 100 mm	0.45 mm	4.11 mm	0.78 mm	4.94 mm	0 mm

When a tensile force is applied to a yarn, due to the yarn shrinking and the out-of-plane compression between overlapped yarns, the fabric thickness decreases and the transverse compressive modulus of the yarn increases. However, the variations of thickness and transverse compressive modulus are difficult to be measured experimentally. According to the thickness-pressure relation shown in Figure 9, Ivanov and Lomov [25] suggested that the thickness variation and the transverse compressive of a yarn can be approximated as bilinear functions of the compressive force. Please note that, although a fabric is subjected to a tension force along the yarn direction, the yarn is actually compressed in the transverse direction due to the yarn shrinking and the out-of-plane compression between overlapped yarns. Thus, the transverse stress state of a yarn in this work is similar to the transvers compressive test shown in the work of Ivanov and Lomov [25]. Besides, the materials involved in Figures 8 and 9 are both carbon fabrics with a 12 K yarn bundle. Thus, the quasi-bilinear trend shown in Figure 9 is reasonably applicable for the validation in this work.

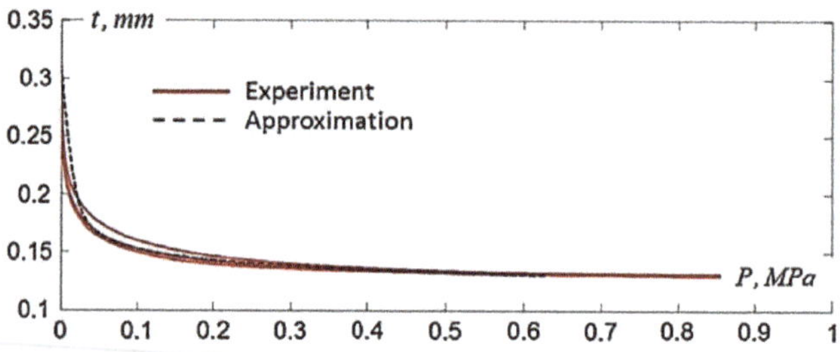

Figure 9. Thickness-compressive pressure of a dry carbon yarn [23].

For the cases in Nosrat-Nezami et al. [22], the maximum tensile load is 2.4 N/mm. If the contact force induced by yarn bending is ignored, the tensile load of 2.4 N/mm corresponds to a compressive pressure of about 0.05 MPa, according to Equations (16) and (18), which is located around the first linear segment in Figure 9. Thus, it is rational to assume that the thickness and transverse compressive moduli linearly vary with the tensile force. In addition, the bending stiffness also varies due to the yarn shrinking. Furthermore, the bending stiffness directly affects the initial contact force F_0^c induced by yarn waviness. Thus, it is also reasonable to set F_0^c as a linear function of the tension force. Then, the thickness t, the compressive modulus E_{com}, and the initial contact force F_0^c can be obtained by a linear curve-fitting process from the cases of no tension and 2.4N/mm tension. Besides, the shear angle when a yarn starts to rotate, φ_0, can be approximately seen as a constant throughout the load process, since it is not sensitive to the contact force between overlapped yarns. Thus, one curve is enough to fit the φ_0. Overall, the cases of no tension and 2.4 N/mm tension are employed to curve-fit model parameters, and the other cases are used for validation. All the curve-fitted parameters are shown in Table 2.

Table 2. Curve-fitting parameters for a carbon fiber plain-woven fabric.

Loading Cases	No Tension	0.8 N/mm	1.6 N/mm	2.4 N/mm
φ_0 (°)	1.5°	1.5°	1.5°	1.5°
F_0^c (N)	0.2	0.33	0.47	0.6
t (mm)	t	0.86 t	0.73 t	0.6 t
E_{com} (MPa)	0.25	0.83	1.41	2

Figure 10 compares the results of the present model and experimental data. The cases of no tension and 2.4 N/mm tension are used to curve-fit the parameters of the present model, t, φ_0, F_0^c, and E_{com}. The cases of 0.8 N/mm and 1.6 N/mm are used for validation. Equation (28) provides a method to quantitatively characterize the average error between experimental data and the predicted results.

$$\mathrm{Err} = \sum_{i=1}^{M} \frac{1}{M} \frac{\sqrt{\left(\tau_{sh}^{pr}(\varphi_i) - \tau_{sh}^{ex}(\varphi_i)\right)^2}}{\tau_{sh}^{ex}(\varphi_i)}, \tag{28}$$

where φ_i is the evenly distributed shear angle point with an angle-interval of around 5°. $\tau_{sh}^{pr}(\varphi_i)$ and $\tau_{sh}^{ex}(\varphi_i)$ are the corresponding predicted and measured shear stresses for each φ_i. M is the number of collected data points of a curve. The average errors for the four load cases are shown in Figure 10. The average errors of the four cases are around 10%, which is good enough considering the scatter of measured data.

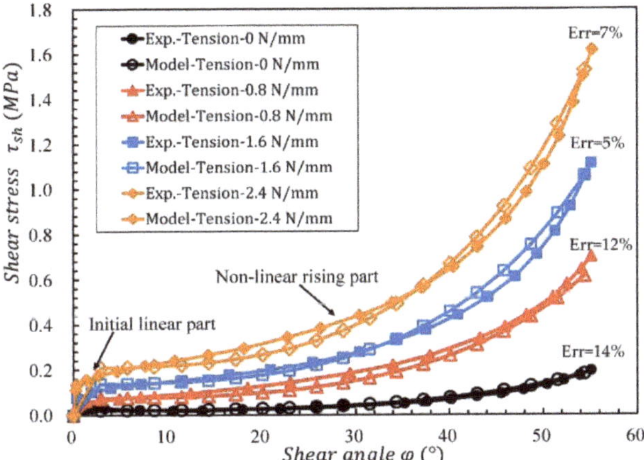

Figure 10. Comparison between the analytical model and experimental data of a carbon plain-woven fabric under bi-axial tension and shear load.

For the no tension and 2.4 N/mm tension cases, the theoretical curves agree well with experimental results, proving that the present model can effectively describe the shear mechanism and the tension-shear coupling effect of fabrics under biaxial tension. For the cases of 0.8 N/mm and 1.6 N/mm, the deviations between predicted and experimental results are very small, verifying the prediction capability of the present model.

Comparing the curves of the four cases, it is found that the shear rigidity of a fabric significantly increases with the applied tension force. Thus, it is evident that the tension-shear coupling must be taken into account in the material forming process of fabrics. As shown in Figure 10, a stress-strain curve can be divided into two parts: the initial linear part and the nonlinear part. The initial part is induced by the static friction between overlapped yarns before a yarn rotates. The maximum value of the static friction force determines the initial shear stress value of the nonlinear part. A higher tension force leads to a larger maximum static friction force, thereby resulting in stiffer shear property of a fabric. Thus, the friction between overlapped yarns is one source of the tension-shear coupling. For the nonlinear part, the stiffness rapidly increases when the tension force varies from zero to 2.4 N/mm. This happens because the fiber density of a yarn becomes larger due to yarn shrinking induced by a tension force. A higher fiber density means a higher transverse compressive Young's modulus. Furthermore, higher transverse stiffness of yarns brings a larger shear resistance for the shear deformation of fabrics. Obviously, the transverse compression of yarns is one main source of the tension-shear coupling.

For general fabrics, there should be a flat part between the initial linear part and the nonlinear rising part. The flat part is induced by friction during shear deformation before transverse compression of parallel yarns occurs. However, the fill gap is zero for the case in Figure 10, meaning that parallel yarns contact each other when a yarn starts to rotate. Thus, no obvious flat part is observed in Figure 10.

4. Parametric Study

It is essential to understand the effect of geometry and mechanical properties of yarns on the large shear behavior. In this work, a kind of carbon fiber woven fabric is taken as an example. It should be noted in Figure 10 that the unit of tension force along a yarn is N/mm, meaning that the force is homogenized by the side length of an RVE. Thus, the force along a yarn is dependent on the thickness of a fabric. Conversely, the parametric study requires that the tensile load is independent of thickness variation. Thus, a tensile stress σ_y with unit of MPa is used. As a reference, some basic parameters are reasonably set in Table 3. The sensitivity study is performed by varying one of these parameters while keeping the others constant.

Table 3. Parameter values of a reference-woven fabric.

Parameters	Value	Parameters	Value
Weave pattern	plain	Initial shear angle φ_0	2°
Yarn width w_f & w_w	5 mm	Tension stress along yarns σ_y	5 MPa
Yarn gap g_f & g_w	0.5 mm	Friction factor μ	0.6
Yarn thickness t_f & t_w	0.5 mm	Compressive modulus of yarn E_{com}	1 MPa

Figure 11a shows the shear deformation curves of fabrics with different weave patterns. The weave pattern is characterized by η_y/N_y, the ratio between the inflection points of a yarn and the number of yarns in an RVE. The ratios are 1, 1/2, and 1/4 for plain woven, 2*2 twill, and 8 harness satin, respectively. As shown in Figure 11a, a higher ratio corresponds to a higher shear stiffness. The result is reasonable because a higher ratio means more interaction between fill and warp yarns.

Figure 11b is the parametric study of the fabric geometry. It is found that the shear stiffness increases with the increase of yarn thickness but decreases with the increase of yarn gap. This is because the waviness angle becomes larger when a yarn becomes thicker. Thus, the contact force between overlapped yarns increases, though the tension force along the yarn remains unchanged. However, the increase of yarn gap reduces the waviness angle, thereby weakening the effect of the tension force on shear deformation. Besides, the shear-locking phenomenon is postponed with the increase of gap size, 25° for $g_0 = 0.5$, 34° for $g_0 = 1$, and 44° for $g_0 = 2$. The transverse compression of parallel yarns is one main source of the shear resistance. Thus, the rising shear-locking angle significantly reduces the shear rigidity of a fabric.

Figure 11c illustrates the effect of the yarn property and the tensile load on the shear behavior of a fabric. It should be noted that in real cases, the transverse compressive modulus E_{com} is sensitive to the tensile load F_t. However, in the sensitivity study, the two parameters are assumed to be independent so as to investigate the effect of each parameter on the shear behavior. It can be seen that the increase of tensile force along yarns results in a larger shear rigidity of a fabric. Moreover, each curve in Figure 11 can be divided into three parts, the initial linear part, the flat part, and the nonlinear rising part. Curves in Figure 11a–c indicate that the weave pattern, the yarn geometry, and the tensile load along yarns can only influence the maximum stress of the first part and the shear stress of the flat part but have no effect on the third part. However, as shown in Figure 11c, the growth of E_{com} significantly increases the slope of the third part of a curve, thereby enhancing the shear rigidity of a fabric. This happens because the effects of the weave pattern, the geometry, and the tensile load on shear behavior are characterized by the friction force which controls the linear and the flat part. However, the third nonlinear rising part is mainly controlled by the transverse compression of yarns.

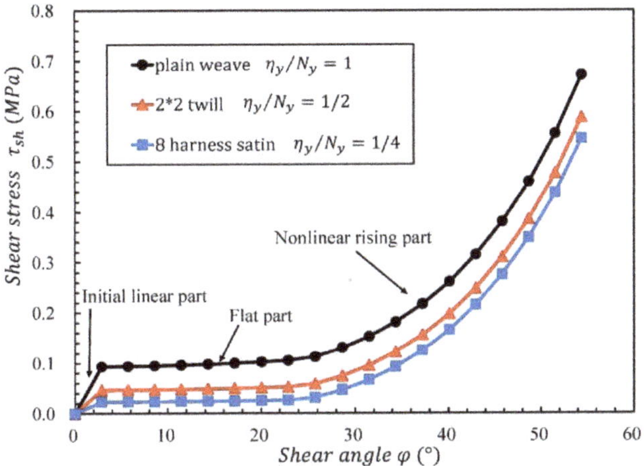

(a) Effect of weave pattern

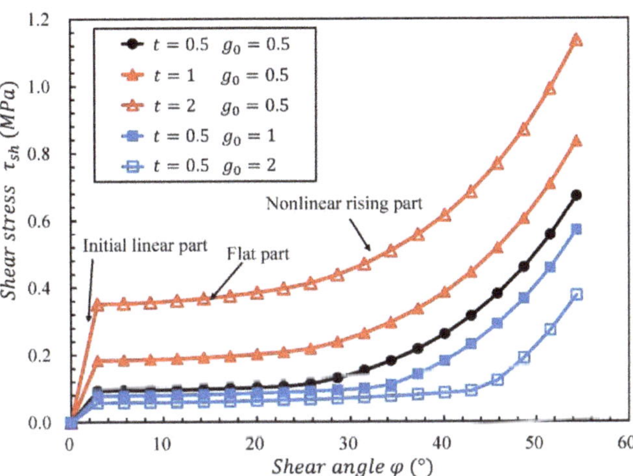

(b) Effect of geometry of yarns

Figure 11. *Cont.*

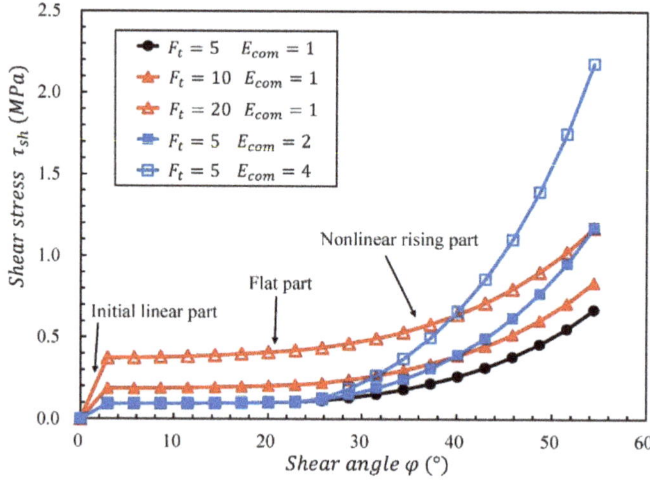

(**c**) Effect of mechanical property and tensile load

Figure 11. Sensitivity study on the shear behavior of a fabric.

5. Conclusions

An analytical model is proposed to describe the tension-shear coupling effect of fabrics with different weave patterns under large shear deformation. Different weave patterns are characterized by the number of inflection points of yarns. In this work, the tension-shear coupling mechanism can be revealed from two aspects, the friction dissipation energy and the in-plane transverse compressive dissipation energy. On the one hand, the subjected tensile force along yarns increases the contact force between overlapped yarns, leading to the growth of the friction dissipation energy. Then, larger friction dissipation energy brings more resistance with respect to shear deformation, thereby increasing the shear rigidity of a fabric. On the other hand, the tensile load along a yarn makes the yarn shrink. The shrinking improves the transverse compressive modulus of the yarn. Besides, the transverse compressive modulus of a yarn is also improved due to the compressive force between overlapped yarns induced by the tension force. Then, the dissipation energy of the transverse compression of yarns becomes larger due to the increase of the mechanical properties of yarns, thus enhancing the shear rigidity of a fabric. In addition to the description of the tension-shear coupling mechanism, the present model has a high computational efficiency due to its explicit and analytical expressions.

The model is validated by experimental results of the shear behavior of a fabric under different biaxial tensile loads. Furthermore, parametric studies are carried out using the present model. In general, fewer inflection points, a larger gap-to-width ratio, and softer in-plane compressive moduli decrease the shear stiffness of a fabric, which can improve the formability of a fabric. When tensile force along yarns is utilized to reduce wrinkling, one should consider that the tensile force will also affect the shear property of a fabric and will further affect the yarn re-orientation. With the help of the present analytical model, engineers can easily optimize the design of fabrics with different weave patterns. Because of the high computational efficiency of the model, the optimization process can be significantly improved.

Author Contributions: Conceptualization, Y.W, Z.H., F.G., Y.L., and Z.Z.; Funding acquisition, Y.W., Z.H., and Z.Z.; Methodology, Y.W., W.Z., and H.R.; Project administration, Z.H. and Z.Z.; Supervision, Z.H., F.G, Y.L., and Z.Z.; Validation, Y.W., W.Z., and H.R.; Writing—original draft, Y.W.; Writing—review & editing, Y.W., W.Z., and H.R. All authors have read and agreed to the published version of the manuscript.

Funding: This research was funded by the U.S. Department of Energy, grant number DE-EE0006867, Chinese Scholarship Council, grant number 201506260043, National Natural Science Foundation of China, grant number 11832014, and Science and Technology projects of Guangdong Province, grant number 2019B090911003.

Acknowledgments: Wangyu Liu and Dong Chen are acknowledged for their constructive suggestions.

Conflicts of Interest: The authors declare no conflict of interest.

References

1. Soleimani, S.M.; Sayyar Roudsari, S. Analytical study of reinforced concrete beams tested under quasi-static and impact loadings. *Appl. Sci.* **2019**, *9*, 2838. [CrossRef]
2. Rampini, M.C.; Zani, G.; Colombo, M.; di Prisco, M. Mechanical behaviour of TRC composites: Experimental and analytical approaches. *Appl. Sci.* **2019**, *9*, 1492. [CrossRef]
3. Poppe, C.; Dorr, D.; Henning, F.; Luise, K. Experimental and numerical investigation of the shear behaviour of infiltrated woven fabrics. *Compos. Part A* **2018**, *114*, 327–337. [CrossRef]
4. Rashidi, A.; Milani, A.S. A multi-step biaxial bias extension test for wrinkling/de-wrinkling characterization of woven fabrics: Towards optimum forming design guidelines. *Mater. Des.* **2018**, *146*, 273–285. [CrossRef]
5. Nosrat Nezami, F.; Gereke, T.; Cherif, C. Active forming manipulation of composite reinforcements for the suppression of forming defects. *Compos. Part A* **2017**, *99*, 94–101. [CrossRef]
6. Harrison, P.; Abdiwi, F.; Guo, Z.; Potluri, P.; Yu, W.R. Characterising the shear–tension coupling and wrinkling behaviour of woven engineering fabrics. *Compos. Part A* **2012**, *43*, 903–914. [CrossRef]
7. Syerko, E.; Comas-Cardona, S.; Binetruy, C. Models for shear properties/behavior of dry fibrous materials at various scales: A review. *Int. J. Mater. Form.* **2015**, *8*, 1–23. [CrossRef]
8. Boisse, P.; Hamila, N.; Guzman-Maldonado, E.; Madeo, A.; Hivet, G.; Dell Isola, F. The bias-extension test for the analysis of in-plane shear properties of textile composite reinforcements and prepregs: A review. *Int. J. Mater. Form.* **2017**, *10*, 473–492. [CrossRef]
9. Cao, J.; Akkerman, R.; Boisse, P.; Chen, J.; Cheng, H.S.; de Graaf, E.F.; Gorczyca, J.L.; Harrison, P.; Hivet, G.; Launay, J.; et al. Characterization of mechanical behavior of woven fabrics: Experimental methods and benchmark results. *Compos. Part A* **2008**, *39*, 1037–1053. [CrossRef]
10. Boisse, P.; Hamila, N.; Vidal-Sallé, E.; Dumont, F. Simulation of wrinkling during textile composite reinforcement forming. Influence of tensile, in-plane shear and bending stiffnesses. *Compos. Sci. Techcol.* **2011**, *71*, 683–692. [CrossRef]
11. Komeili, M.; Milani, A.S. On effect of shear-tension coupling in forming simulation of woven fabric reinforcements. *Compos. Part B* **2016**, *99*, 17–29. [CrossRef]
12. Zhang, W.Z.; Bostanabad, R.; Liang, B.; Su, X.M.; Zeng, D.; Bessa, M.A.; Wang, Y.C.; Chen, W.; Cao, J. A numerical Bayesian-calibrated characterization method for multiscale prepreg preforming simulations with tension-shear coupling. *Compos. Sci. Techcol.* **2019**, *170*, 15–24. [CrossRef]
13. Robitaille, F.; Clayton, B.R.; Long, A.C.; Souter, B.J.; Rudd, C.D. Geometric modelling of industrial preforms: Woven and braided textiles. *Part L J. Mater. Des. Appl.* **1999**, *213*, 69–83. [CrossRef]
14. Long, A.C.; Rudd, C.D.; Blagdon, M.; Smith, P. Characterizing the processing and performance of aligned reinforcements during preform manufacture. *Compos. Part A* **1996**, *27*, 247–253. [CrossRef]
15. King, M.J.; Jearanaisilawong, P.; Socrate, S. A continuum constitutive model for the mechanical behavior of woven fabrics. *Int. J. Solids Struct.* **2005**, *42*, 3867–3896. [CrossRef]
16. Peng, X.Q.; Cao, J. A continuum mechanics-based non-orthogonal constitutive model for woven composite fabrics. *Compos. Part A* **2005**, *36*, 859–874. [CrossRef]
17. Gong, Y.; Yan, D.X.; Yao, Y.; Wei, R.; Hu, H.L.; Xu, P.; Peng, X.Q. An anisotropic hyperelastic constitutive model with tension–shear coupling for woven composite reinforcements. *Int. J. Appl. Mech.* **2017**, *9*, 1750083. [CrossRef]
18. Lomov, S.V.; Verpoest, I. Model of shear of woven fabric and parametric description of shear resistance of glass woven reinforcements. *Compos. Sci. Techcol.* **2006**, *66*, 919–933. [CrossRef]
19. Zhu, B.; Yu, T.X.; Teng, J.; Tao, X.M. Theoretical modeling of large shear deformation and wrinkling of plain woven composite. *J. Compos. Mater.* **2009**, *43*, 125–138. [CrossRef]
20. Erol, O.; Powers, B.M.; Keefe, M. A macroscopic material model for woven fabrics based on mesoscopic sawtooth unit cell. *Compos. Struct.* **2017**, *180*, 531–541. [CrossRef]

21. Nasri, M.; Garnier, C.; Abbassi, F.; Labanieh, A.R.; Dalverny, O.; Zghal, A. Hybrid approach for woven fabric modelling based on discrete hypoelastic behaviour and experimental validation. *Compos. Struct.* **2019**, *209*, 992–1004. [CrossRef]
22. Nosrat-Nezami, F.; Gereke, T.; Eberdt, C.; Cherif, C. Characterisation of the shear–tension coupling of carbon-fibre fabric under controlled membrane tensions for precise simulative predictions of industrial preforming processes. *Compos. Part A* **2014**, *67*, 131–139. [CrossRef]
23. Najjar, W.; Pupin, C.; Legrand, X.; Boude, S.; Soulat, D.; Dal Santo, P. Analysis of frictional behaviour of carbon dry woven reinforcement. *J. Reinf. Plast. Compos.* **2014**, *33*, 1037–1047. [CrossRef]
24. Allaoui, S.; Hivet, G.; Wendling, A.; Ouagne, P.; Soulat, D. Influence of the dry woven fabrics meso-structure on fabric/fabric contact behavior. *J. Compos. Mater.* **2012**, *46*, 627–639. [CrossRef]
25. Ivanov, D.S.; Lomov, S.V. Compaction behaviour of dense sheared woven preforms: Experimental observations and analytical predictions. *Compos. Part A* **2014**, *64*, 167–176. [CrossRef]

Article

Numerical Analysis of the Influence of Empty Channels Design on Performance of Resin Flow in a Porous Plate

Glauciléia M. C. Magalhães [1], Cristiano Fragassa [2,*], Rafael de L. Lemos [3], Liércio A. Isoldi [1,3], Sandro C. Amico [4], Luiz A. O. Rocha [5], Jeferson A. Souza [1,3] and Elizaldo D. dos Santos [1,3]

1 Graduate Program of Computational Modeling, Federal University of Rio Grande, Italia Avenue, km 8, Rio Grande 96201-900, Brazil; mglaucileia@gmail.com (G.M.C.M.); liercioisoldi@furg.br (L.A.I.); jefersonsouza@furg.br (J.A.S.); elizaldosantos@furg.br (E.D.d.S.)
2 Interdepartmental Center for Industrial Research on Advanced Applications in Mechanical Engineering and Materials Technology, University of Bologna, Viale Risorgimento 2, 40136 Bologna, Italy
3 Graduate Program of Ocean Engineering, Federal University of Rio Grande, Avenue, km 8, Rio Grande 96201-900, Brazil; er.lemos@outlook.com
4 Materials Engineering Graduate Program, University of Rio Grande do Sul, Bento Gonçalves Av., 9500, Porto Alegre 91501-970, Brazil; amico@ufrgs.br
5 Mechanical Engineering Graduate Program, University of Vale do Rio dos Sinos, São Leopoldo 93022-750, Brazil; luizor@unisinos.br
* Correspondence: cristiano.fragassa@unibo.it; Tel.: +39-347-697-4046

Received: 19 March 2020; Accepted: 17 April 2020; Published: 11 June 2020

Featured Application: The influence of empty channel design on resin flow in a porous plate is studied. Constructal design is used for geometric investigation of I and T-shaped channels. The multiphase resin/air flow in the porous plate is solved numerically. T-shaped configurations with long branches lead to the best impregnation performance. Permanent air voids arise in the domain for very thin T-shaped channels.

Abstract: This numerical study aims to investigate the influence of I and T-shaped empty channels' geometry on the filling time of resin in a rectangular porous enclosed mold, mimicking the main operating principle of a liquid resin infusion (LRI) process. Geometrical optimization was conducted with the constructal design (CD) and exhaustive search (ES) methods. The problem was subjected to two constraints (areas of porous mold and empty channels). In addition, the I and T-shaped channels were subjected to one and three degrees of freedom (DOF), respectively. Conservation equations of mass and momentum for modeling of resin/air mixture flow were numerically solved with the finite volume method (FVM). Interaction between the phases was considered with the volume of fluid method (VOF), and the effect of porous medium resistance in the resin flow was calculated with Darcy's law. For the studied conditions, the best T-shaped configuration resulted in a filling time nearly three times lower than that for optimal I-shaped geometry, showing that the complexity of the channels benefited the performance. Moreover, the best T-shaped configurations were achieved for long single and bifurcated branches, except for configurations with skinny channels, which saw the generation of permanent voids.

Keywords: constructal design; resin flow; porous media; numerical simulation; filling time

1. Introduction

Liquid composite molding (LCM) processes were developed to enable the production of large, wide, and complex structural components at low cost. Among the important LCM processes, it is important to mention liquid resin infusion (LRI), which consists of the injection of a polymer resin through empty channels mounted on a fibrous mold, favoring the global propagation of the resin over the mold domain [1]. One of the most considerable difficulties in applying LRI is related to complete mold filling and void formation, ensuring that the fibrous reinforcement is completely impregnated by the resin within the mold [2–5].

To tackle the reported difficulties, several efforts have been made to improve process comprehension, as well as to avoid traditional trial-and-error approaches. One possibility is the use of computational tools to simulate multiphase resin/air flow into the enclosed mold [3–8]. The study of parametrical investigations in the LRI process (as the influence of design over the behavior of resin/air flow, voids formation, and filling time) also represents an important subject. Therefore, studies have been carried out in the analytical, experimental, and numerical scope to improve the understanding of resin/air flow in this kind of domain [3,4].

The development of computational methods for the representation of LCM processes is an important point of concern. For example, Isoldi et al. [9] performed a numerical study of resin flow in porous domains that mimics the operational principle of the Resin Transfer Molding (RTM) and Light Resin Transfer Molding (LRTM) processes. Different resin flow cases were solved numerically (linear, radial, and complex three-dimensional domain) and compared with analytical, experimental, and numerical results available in the literature. The maximum difference between these solutions was about 8.0%, showing the validity of the developed numerical method. Wang et al. [1] conducted a numerical and experimental study of the resin infusion process under industrial conditions. The authors verified that mold filling time determined by numerical simulation brought about results in agreement with experimental solutions. Afterward, Grössing et al. [10] compared the technical feasibility of using two different numerical methods (PAM-RTM® and OpenFOAM) to obtain numerical solutions of resin flow. Firstly, it was found that the numerical methods led to valid solutions in comparison with experimental results. It was also verified that both numerical methods had very similar results for the simulation of resin flow in porous media. Sirtautas et al. [11] developed a numerical model based on the Finite Element Method (FEM), considering the effects of compressibility and orthotropic permeability dependent on pressure in the domain for different fibrous materials. A comparison between the numerical model and experimental results was performed for resin flow in the two-dimensional domain of an aerospace piece. The results showed good agreement, recommending the use of the proposed computational method. Pierce et al. [12] developed and validated a multi-physics model to improve the simulation of the infusion process in the complex preform. The results are compared with traditional models that do not consider fabric deformation. The combination of deformation-dependent permeability properties and the preform draping model led to more realistic predictions for local infusion flow. Chebil et al. [13] developed a computational model for the simulation of three-dimensional resin flow in laminated preform composed of multiple layers with different permeabilities. The method proposed the use of shell elements for flow simulation instead of solid elements, leading to a reduction in computing time. Rubino and Carlone [14] proposed an analytical/numerical methodology to consider the effects of preform compaction on resin flow for a deformable porous media, which is important for processes such as vacuum assisted resin transfer molding (VARTM) and Seeman composites resin infusion molding process (SCRIMP), where a flexible bag is generally used as part of the mold. The results showed that preform compaction affected resin flow and filling time.

The numerical approach has also been used to define strategies for several LCM processes. Examples of the evaluation of gate placements in RTM processes seeking to prevent void content and minimize filling time had been widely reported in different works [15–20]. The work of Brouwer et al. [3] proposed strategies to improve the positioning of injection channels in one LRI process using computational modeling. More precisely, the authors presented developments in two large structures to be filled with resin, a rotor blade, and a boat hull 20 m and 16 m long, respectively. The strategy adopted by the authors consisted of a main injection channel in the central region of the domain (the keel for the case of injection in the hull domain) with several bifurcated channels spread along the area of the domain (from the center to the periphery). Afterward, Gomes et al. [21] performed a numerical and experimental study for better characterization of a VARTM (vacuum assisted resin transfer molding) process, indicating the best strategies for the filling of a C-shaped stringer. The results indicated that the consideration of two exit regions (with vacuum) and one entering region of resin placed in the extremes of domain length was the best strategy to minimize filling time and prevent void formation. Wang et al. [1] developed an iterative method based on a centroidal Voronoi diagram for optimization of the distribution of several channels to reduce the number of required tests.

Recently, Struzziero and Skordos [22] performed a numerical multi-objective optimization focusing on the choice of an optimal temperature profile, which minimizes the filling time and the risk of hindering resin flow due to excessive curing. For the optimization, the authors employed the genetic algorithm (GA) method. It achieved reductions of nearly 66% and 15% in filling time and final degree of cure, respectively, in comparison with standard solutions. Geng et al. [23] investigated the behavior of resin impregnation in curved porous plates mimicking a VARTM process. They performed an experimental study investigating the influence of curvature angles and preform layers on the advancement of the resin front line. The results indicated that due to interaction between the preforms and void formation, the curved regions led to an augmentation of 15% to 30% of filling time in comparison with a linear region. Shevtsov et al. [24] performed an experimental and numerical study for the manufacturing of three-dimensional composite parts of complex shape. For the computational model, a numerical technique based on liquid resin flow into porous preform is coupled with a model that describes the fluid motion in unsaturated soils. The developed method allowed the prediction of inner and outer dry spots depending on the resin flow movement. Despite several important works reported, few studies can be found on the evaluation of the influence of the geometry of empty channels in the behavior of resin/air fluid flow into the porous domain. In the present study, a geometrical optimization of I and T-shaped empty channels over the required time to fill a porous rectangular plate is performed with the constructal design (CD) and exhaustive search (ES) methods.

Constructal design is a method for assessment of design in any finite flow system based on the objectives and constraints principle. This method has been widely applied to show that shape and structure in nature (trees, rivers, animal body, and others) are deterministically generated following a physical principle named Constructal Law of Design and Evolution [25–29]. According to Bejan [28], constructal law states that the design of the finite-size flow system to persist in time (to live) must evolve freely to maximize access to its internal currents. This method has also been applied as a powerful tool to improve the design of several engineering problems as cooling systems, heat exchangers, renewable energy, transport systems, and even solid mechanics [30–36]. As per the authors' knowledge, except for the works of Refs. [37,38], the constructal design method has not been employed in the literature for investigation on the influence of geometry over impregnation of resin in a porous medium, which mimics the LRI process, a novelty in the present work.

Here, the application of computational modeling, constructal design, and exhaustive search (ES) in geometrical optimization of I and T-shaped empty channels inserted into the rectangular porous medium is proposed. The main aims here are to minimize the filling time of resin impregnation in the porous medium, prevent the formation of permanent voids in the domain, and investigate the influence of geometry over the performance indicator of the system. The solution domain is presented in a two-dimensional rectangular plate with a porous medium, which simulates a fibrous reinforcement. This work is a sort of continuity of the study by Magalhães et al. [38] with the difference being that here the T-shaped channel is completely optimized, i.e., all three degrees of freedom are investigated. The conservation equations of mass, momentum, and one equation for the transport of volumetric fractions are solved with the finite volume method (FVM) [39–41]. To deal with the resin/air interface flow, the volume of fluid (VOF) method is used [42]. In the region of the porous medium, it is also considered a body force modeled with Darcy's law. More precisely, the simulations were performed using the computational fluid dynamics code FLUENT [41].

2. Mathematical Modeling

In this study, incompressible, laminar, and transient flow of a resin/air mixture in a two-dimensional domain is considered. Moreover, the two phases are treated as immiscible. For the prediction of this kind of flow, conservation equations of mass and momentum for the mixture of resin/air and one transport equation for prediction of resin volume fraction are numerically solved. The conservation equations of mass and momentum for the resin/air mixture are given by [43]:

$$\frac{\partial \rho}{\partial t} + \nabla \cdot \left(\rho \vec{V} \right) = 0 \tag{1}$$

$$\frac{\partial \left(\rho \vec{V} \right)}{\partial t} + \nabla \cdot \left(\rho \vec{V} \vec{V} \right) = -\nabla P + \nabla \cdot \overline{\overline{\tau}} + \vec{F} \tag{2}$$

where ρ is the mixture density (kg/m^3), $\vec{V}$ is the velocity vector (m/s), t is the time (s), μ is the dynamic viscosity of the fluid (Pa.s), ∇P is the pressure gradient (Pa/m), $\vec{F}$ is an external force vector per unit volume (N/m^3) and $\overline{\overline{\tau}}$ is the stress tensor of the fluid (N/m^2), given by:

$$\overline{\overline{\tau}} = \mu \left(\nabla \vec{V} + \nabla \vec{V}^T \right) \tag{3}$$

The effect of the porous medium is included in the mathematical model by the insertion of a body force in the momentum equation based on Darcy's law, as given by [43–45]:

$$\vec{F} = -\frac{\mu}{K} \vec{V} \tag{4}$$

To tackle the resin/air mixture, the volume of fluid (VOF) method is employed [42]. In this approach, an additional transport equation for resin volume fraction (f) is necessary to define the quantity of resin along with each cell of the domain. This transport equation is given by [42]:

$$\frac{\partial f}{\partial t} + \nabla \cdot \left(f \vec{V} \right) = 0 \tag{5}$$

With the definition of volume fraction, density and dynamic viscosity in each cell of the computational domain can be calculated by [46]:

$$\rho = f \rho_{res} + (1 - f) \rho_{air} \tag{6}$$

$$\rho = f \mu_{res} + (1 - f) \mu_{air} \tag{7}$$

2.1. Problem Description

As previously mentioned, an incompressible, laminar, and transient multiphase flow of resin/air was considered, moving through a two-dimensional plate domain composed of a porous medium, as shown in Figure 1a,b. Empty channels are inserted along the porous domain to facilitate impregnation of the resin throughout the mold. In Figure 1a,b, these channels (I and T-shaped, respectively) are represented by the dark gray region, while the porous region is represented by light gray.

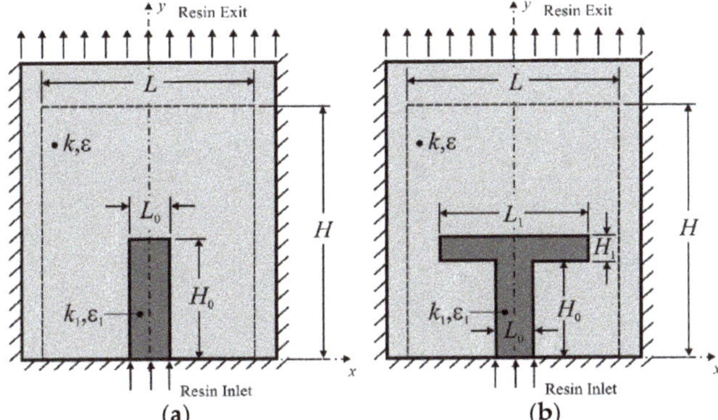

Figure 1. Computational domain of the problem, which mimics the liquid resin infusion (LRI) process with two empty channels geometries: (**a**) I-shaped, (**b**) T-shaped.

In both cases, flow is imposed by a pressure difference between the resin inlet (lower surface of the empty channel) and the exit region of multiphase flow (upper surface of the porous medium). Inlet pressure of $P_{in} = 1 \times 10^5$ Pa is considered, and pressure of $P_{out} = 0$ Pa is imposed in the exit. It is worth mentioning that, in a real LRI process, resin flow is driven by vacuum imposition at the exit of the fibrous reinforcement. Despite this, a pressure difference with the imposition of positive pressure in the resin inlet surface is represented. For computational modeling, it is important to predict the pressure difference. Hence, the imposition made here led to the same results as that achieved when the same pressure difference is imposed with the vacuum in the domain exit. In the remaining domain, bottom, and lateral surfaces, a non-slip and impermeability boundary condition is imposed ($u = v = 0$ m/s). For the resin volume fraction (Equation (5)), $f = 1$ at the inlet section and zero gradients (normal to the surface) on all the other surfaces are imposed. Concerning the physical properties of the resin and air, densities of $\rho_{res} = 916$ kg/m^3 and $\rho_{air} = 1.225$ kg/m^3 and dynamic viscosities of $\mu_{res} = 0.06$ kg/(ms) and $\mu_{air} = 1.7894 \times 10^{-5}$ kg/(ms) are adopted. For the porous media, a permeability of $K = 3.0 \times 10^{-10}$ m^2 and porosity of $\varepsilon = 0.88$ is imposed.

2.2. Geometrical Evaluation

Geometrical optimization was done by using an association between constructal design and exhaustive search, as explained in recent literature [47]. Figure 2 illustrates the main steps employed for optimization of the problem. Steps 1 to 3 presents the flow system and identifies a performance indicator (here, it is filling time). Steps 4 and 5 show the constraints of the problem, degrees of freedom, and the main parameters of the physical problem. It is also necessary to understand the physical problem and define a method for its solution (numerical solution in this case).

With the defining of geometry and understanding of the physical solution method, it is possible to predict the performance indicator (minimization of filling time of resin impregnation along with the porous domain) for each configuration studied. After Step 6, it is necessary to define if the design should be optimized or not.

If the problem is not optimized, a geometrical evaluation is performed, seeking to understand the effect of degrees of freedom over performance indicator, simulating the total number of cases defined. However, if the problem is optimized, it is necessary to specify an optimization method. Here, exhaustive search, which consists of evaluating several geometrical possibilities taking into consideration fixed increments of variation for each degree of freedom (as depicted in Figure 3) for the T-shaped case.

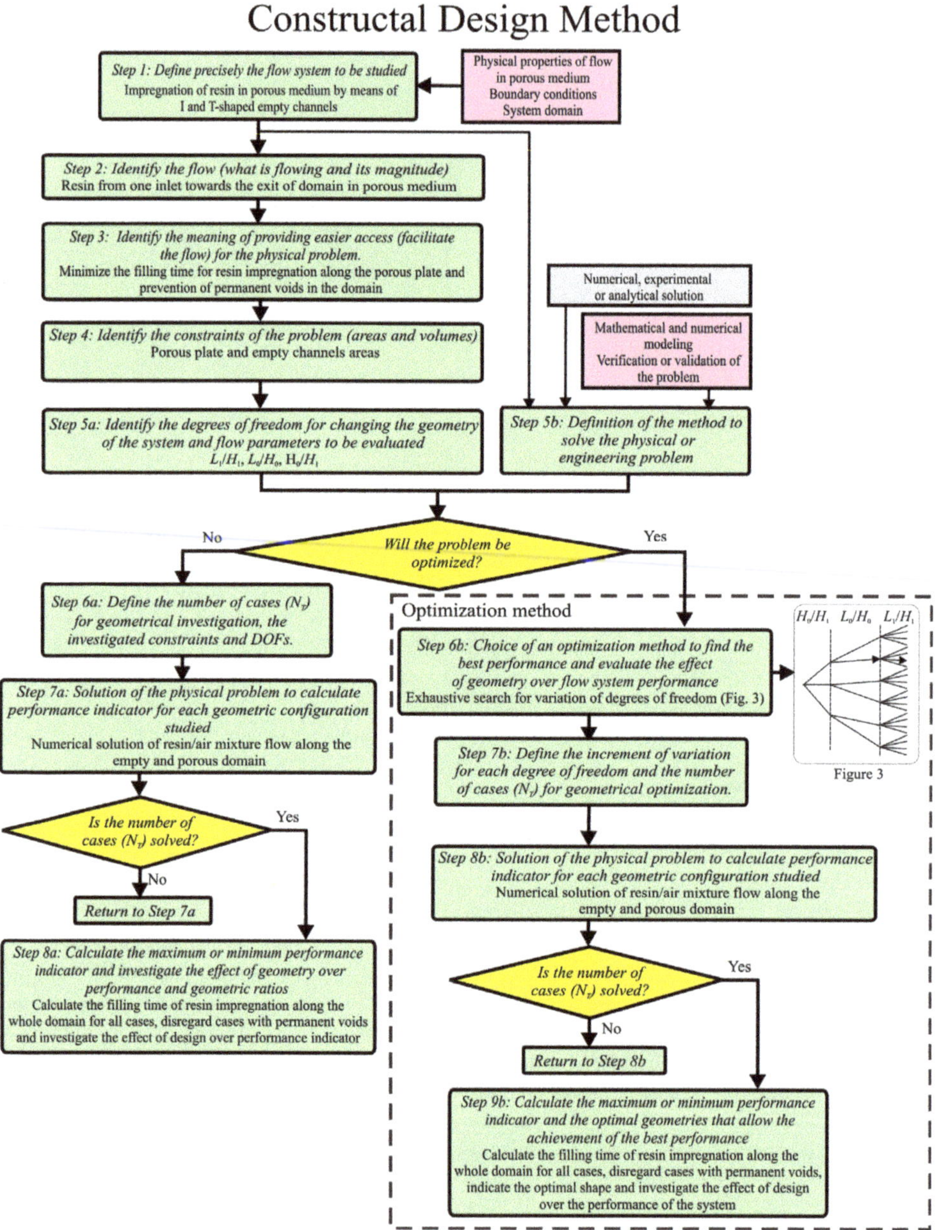

Figure 2. Flow chart illustrating the main steps of the geometrical optimization of I and T-shaped empty channels inserted into the rectangular porous plate for resin impregnation along the domain.

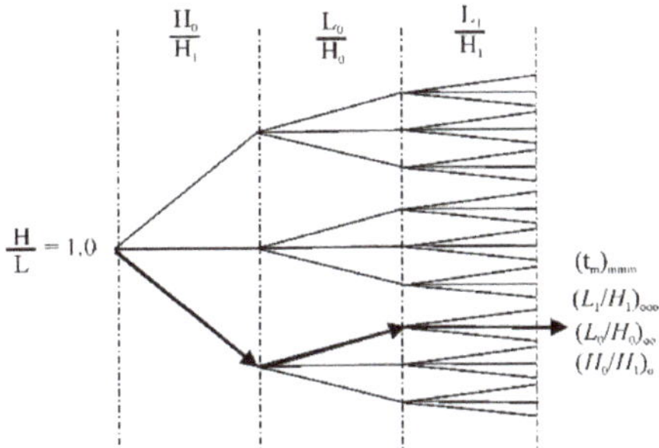

Figure 3. Flowchart of the performed simulations in geometric evaluation of the T-shaped empty channel in resin infusion simulation.

For complete optimization, Steps 6b to 9b, illustrated in Figure 2, are executed. It should be emphasized that heuristic-based optimization methods, e.g., genetic algorithm and simulated annealing, have also been associated with constructal design [48]. These kinds of tools are more used in software packages such as FLUENT [41] and Magma[5] [49].

These methods required a significantly lower number of cases to be solved, but need verification with exhaustive search for suitable reproduction of the effects of DOFs over the performance indicator.

Each case under analysis is submitted to two restrictions, the total area of the computational domain:

$$A = HL \tag{8}$$

and the area of the empty channel, which for I-shaped and T-shaped channels, respectively, are represented by:

$$A_I = H_0 L_0 \tag{9}$$

$$A_T = H_0 L_0 + H_1 L_1 \tag{10}$$

The dimensions of the external domain with porous medium are: $H = L = 0.5$ m for all simulations, i.e., the outer area is a geometric constraint of the problem, and $H/L = 1.0$ is taken into account. It should be noted that a spacing of 0.05 m was taken into consideration for the sides of the fibrous reinforcement mold, and an additional height of 0.25 m was inserted to prevent void regions in the locality defined by the dashed line shown in Figure 1a,b, which is the main analysis region of this study. The areas A_I and A_T can be represented by the ratio between the empty channel and plate areas (fraction area), written by:

$$\phi = \frac{A_I}{A} = \frac{A_T}{A} \tag{11}$$

For the I-shaped empty channel, one degree of freedom (L_0/H_0) is considered, whereas the T-shaped empty channel has three degrees of freedom (L_0/H_0, H_0/H_1, and L_1/H_1). To optimize the geometry of the I-shaped channels, a set of 60 simulations were performed with different L_0/H_0 ratios and different values of ϕ, where $\phi = 0.005$; 0.01; 0.03; 0.05, and 0.1. In the case of T-shaped channels, only $\phi = 0.05$ was evaluated, due to the high number of simulations required for each magnitude of ϕ. In this specific case, a total of 189 simulations were performed. For this case (T-shaped channel), the optimization process was divided into three steps, as shown in Figure 3.

Figure 3 illustrates that in the first optimization step, the ratio L_1/H_1 is varied, while the remaining parameters (L_0/H_0 and H_0/H_1) are kept fixed. To clarify how the terms are assigned, note that the minimum value found for the filling time is named the once minimized filling time (t_m), and the corresponding ratio L_1/H_1 is called the once-optimized ratio of L_1/H_1, $(L_1/H_1)_o$. In a second step, the same process is repeated by varying L_1/H_1 for the other ratios of L_0/H_0 and maintaining H_0/H_1 constant. At this stage of the process, two degrees of freedom are optimized (L_1/H_1 and L_0/H_0). In this case, the minimum filling time obtained is twice minimized (t_{mm}), the L_1/H_1 ratio is the twice-optimized ratio $(L_1/H_1)_{oo}$, and the L_0/H_0 ratio is once optimized $(L_0/H_0)_o$. Finally, the second step is repeated for the different ratios of H_0/H_1 evaluated. Thus, the minimum filling time is the thrice-minimized (t_{mmm}), and the optimal shapes are the thrice-optimized ratio of L_1/H_1, $(L_1/H_1)_{ooo}$, the twice optimized ratio of L_0/H_0, $(L_0/H_0)_{oo}$, and the once-optimized ratio of H_0/H_1, $(H_0/H_1)_o$, completing the geometric optimization process.

3. Numerical Modeling and Code Validation

For simulation of the infusion process, the conservation equations of mass, momentum, and transport equation of volume fraction were solved using a commercial CFD (computational fluid dynamics) code based on FVM [39–41]. Numerical simulations were carried out using computers with 4 Intel Xeon processors and 8 threads of 3.30 GHz clock and 8.0 Gb of RAM. In all the simulations, the second-order upwind discretization scheme was used for the treatment of the advective terms. For pressure interpolation, the PRESTO scheme was used. Pressure-velocity coupling was performed with the PISO method, while the Geo-Reconstruction method was used to reconstruct the interface between the two fluids. In addition, sub-relaxation factors of 0.3 and 0.7 are imposed for the conservation equations of mass and momentum, respectively. The simulations are considered converged when the residuals of conservation equations of mass and momentum are lower than $R < 10^{-6}$.

For time discretization, a study of time step independence was performed, since a transient problem was being investigated. Five simulations were performed with a varying time step, and its influence over the volume of domain completely impregnated by the resin was investigated, i.e., the volume of domain with volume fraction of $f = 1.0$. For all simulations, the volume of resin injected up to time of $t = 7.0$ s was evaluated. Table 1 shows, for the analyzed injection time, the volume percentage of the domain filled with resin and the processing time required to carry out the simulations. The analyzed case consists of a I-shaped empty channel with area fraction of $\phi = 0.05$, $H/L = 1.0$ and $L_0/H_0 = 4.0$. It is possible to observe that for values of $\Delta t \leq 1.0 \times 10^{-3}$ s, all employed time steps lead to identical results. For the values of $\Delta t = 1.0 \times 10^{-2}$ s and $\Delta t = 1.0 \times 10^{-1}$ s, there is no convergence in the simulations. Consequently, the time step of $\Delta t = 1.0 \times 10^{-3}$ s is adopted in subsequent geometric evaluation simulations, as it leads to a lower computational effort compared to smaller time steps.

Table 1. Study of sensibility of time step over resin impregnation for an I-shaped empty channel with $\phi = 0.05$, $H/L = 1$, and $L_0/H_0 = 4$.

Time Step (s)	Mold Filling (%)	Processing Time (s)
1.0×10^{-5}	27.1596	171,000
1.0×10^{-4}	27.1596	8900
1.0×10^{-3}	27.1596	1200
1.0×10^{-2} *	—	—
1.0×10^{-1} *	—	—

* Non-converging solutions.

For spatial discretization, the domain was divided into finite rectangular volumes, and a mesh independence test was performed to define the number of volumes used for the simulations. The time step used is defined in Table 1 ($\Delta t = 1.0 \times 10^{-3}$ s). Table 2 shows the number of volumes, the filling time for complete impregnation of resin in the mold, and the processing time required to perform the simulations for each evaluated mesh and for the same I-shaped channel studied in the time independence study, i.e., $\phi = 0.05$, $H/L = 1.0$ and $L_0/H_0 = 4.0$. The following equation presents the criterion for achievement of an independent mesh:

$$Difference = \frac{100 \cdot \left|t^j - t^{j+1}\right|}{t^j} < 1.0\% \tag{12}$$

where t^j represents the minimum value of the filling time calculated with the coarser mesh, and t^{j+1} corresponds to the magnitude calculated with the refined successive mesh. Successive refinements determine appropriate mesh size until the magnitude of the difference variable for two successive meshes is lower than 1.0%. In this sense, a mesh with 18,271 finite rectangular volumes is used in this study.

Table 2. Mesh independence test considering an I-shaped empty channel with $\phi = 0.05$, $H/L = 1.0$ and $L_0/H_0 = 4.0$.

Number of Volumes	Filling Time (s)	Difference (%)	Processing Time (s)
1209	188.3	1.69	43,000
4636	185.1	1.18	78,400
18,271	182.9	0.87	126,500
72,451	181.3	—	184,900

To show the reliability of the present computational model, a verification of the numerical method used here was done by comparing the present numerical solution and a classical analytical solution for the rectilinear case. More precisely, resin advancement as a function of infusion time obtained with the present method and the analytical solution described in Jinlian et al. [50] and Rudd [45] are compared. This comparison can be made for empty channels that lead to rectilinear behavior of resin in the porous medium. For the sake of comparison, a simulation in an I-shaped empty channel with $\phi = 0.05$, $H/L = 1.0$, and $L_0/H_0 = 19.0$, where L_0 is equal to L, is chosen. The case with these configurations represents a rectilinear case.

An analytical solution for front-line resin advancement as a function of time is given by [50]:

$$X_f = \sqrt{\frac{2KP_{in}t}{\mu\varepsilon}} \tag{13}$$

where X_f is the position of the resin front-line (m), t is the time (s), μ is the resin viscosity (Pa/s), ε is the porosity, K is the permeability (m^2), and P_{in} is the injection pressure (Pa). It should be noted that the present formulation is only applicable to a constant P_0 situation.

A monitoring line was created in the center of the domain to obtain the numerical results. More precisely, a line that overlaps the y-axis was defined by the following points: P$_1$ ($x_1 = 0.0$ m, $y_1 = 0.0256$ m) and P$_2$ ($x_2 = 0.0$ m, $y_2 = 0.7500$ m). The coordinates $x_1 = x_2$ represent the center of the domain in the x-direction, while the coordinate y_1 represents the interface placement between the open channel and the porous medium, and y_2 is the exit region of the resin. Figure 4 shows the advancement of the resin front line as a function of the injection time for both solutions (analytical and numerical) and an illustration of the simulated domain with monitoring points for resin advancement in the porous medium. A maximum difference of 0.66% was observed between the analytical and numerical results, which is an indication of very good agreement. Therefore, the verification of the numerical model presented was satisfactorily achieved, and the code can be used for theoretical recommendations on the geometric configurations of I and T-shaped empty channels inserted in a rectangular porous plate.

It is also worth mentioning that the present computational model was previously verified and validated in the work by the authors of [9]. The resin propagation obtained with this computational model was verified with the analytical data for linear and radial cases and with numerical results obtained with the PAM-RTM® software for three-dimensional resin flow in a rectangular box and spherical shell. Moreover, the code was validated with the experimental results obtained by Schmidt et al. [51] for a case of resin flow in a cavity mold with an inlet nozzle. More precisely, it was compared to the results of resin flow advancement as a function of time: differences of nearly 8.0% were found. The cases in Reference [9] mimic resin transfer molding (RTM) or light-resin transfer molding (LRTM) processes. In spite of the simulation of different processes, the fluid dynamic behavior of the resin flow of Reference [9] is the same as performed here. These results were reproduced for code validation in this study, but are not provided here since they have already been presented in Reference [9].

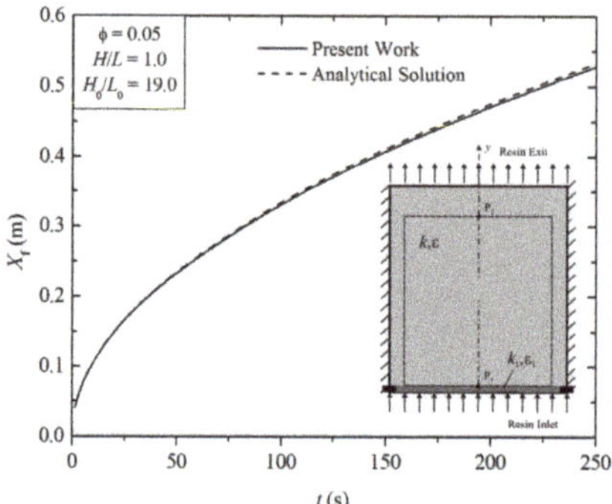

Figure 4. Comparison between the placement of resin flow as a function of time by analytical solution.

4. Results and Discussion

To eliminate residual air from a porous medium, it is necessary to perform successful molding. The formation of voids due to partial impregnation of resin inside the fibrous reinforcement can lead to failures in the use of the final product [45]. In this sense, geometric optimization is essential because it not only increases productivity by reducing the time of injection, but also enhances the quality of the pieces produced. Thus, the prevention of the formation of voids in the porous medium is a restriction in the analysis and definition of the liquid resin infusion (LRI) process.

A reference line was created in the domain to measure filling time (see the horizontal dashed lines in Figure 1a,b, defined by the following points: P_1 ($x_1 = -0.25$ m, $y_1 = 0.50$ m) and P_2 ($x_2 = 0.25$ m, $y_2 = 0.50$ m)). The position of the front line of the resin flow is monitored along this line. When the resin flow completely crosses the reference line, i.e., when the volumetric fraction along the entire line is equal to $f = 1.0$, the infusion process is considered complete.

To evaluate the geometric influence of the I-shaped empty channel over the filling infusion time of the resin along the rectangular porous plate, the fraction area inside the domain is analyzed. Simulations performed with different L_0/H_0 ratios are used to determine the case with lower infusion time for five different fraction areas of the empty channel, $\phi = 0.005$; 0.01; 0.03; 0.05 and 0.1. It has been found that the effect of the ratio L_0/H_0 over filling time (t) is similar for all investigated fractions ϕ. The only exception happened for $\phi = 0.1$ due to the restriction of the geometry in the upper limit of L_0/H_0, as can be seen in Figure 5. In general, it is noticed that the lowest L_0/H_0 ratios lead to the best results, i.e., there is a reduction in filling time (t) when the channel has higher penetration in the y-direction of the domain. However, the progressive increase of L_0/H_0 to the upper limit results in a decrease in injection time too. Thus, there is a globally optimized ratio placed in the lowest region of L_0/H_0 and a local optimized ratio region, placed in the superior limit of the ratio L_0/H_0 investigated. The worst performance, for almost all cases of ϕ, is reached for intermediate ratios of L_0/H_0. Different behavior is noticed only for $\phi = 0.100$, where there is an almost asymptotic growth of filling time as a function of ratio L_0/H_0, showing that restriction influences the best shape and effect of geometry over the performance indicator.

The results also showed that the best configurations for lower magnitudes of ϕ can have a superior performance than those achieved for intermediate ratios of L_0/H_0, even for higher magnitudes of ϕ. For example, the best configuration obtained for $\phi = 0.01$, $(L_0/H_0)_o = 0.015$ led to a filling time of $t_m = 105.3$ s, which is lower than the filling time reached for several configurations (not optimized) with $\phi = 0.03$; 0.05 and 0.1. This behavior indicates that geometric rationalization can lead to smaller dimension channels for a performance superior to that reached for larger channels, where geometric configuration is not optimized.

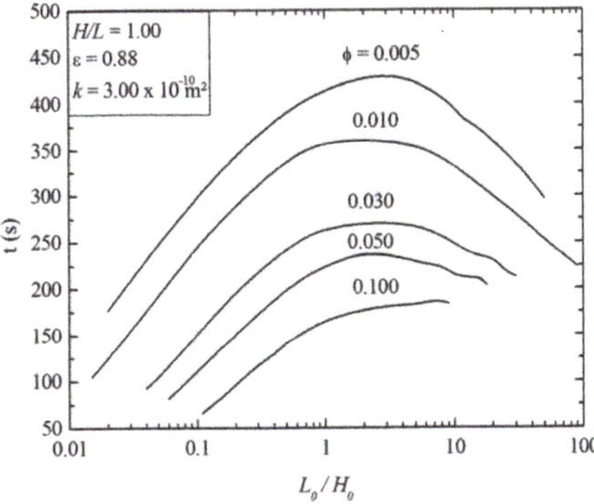

Figure 5. Effect of the L_0/H_0 ratio over resin infusion filling time for different area fractions of the I-shaped empty channels.

The optimal results obtained in Figure 5 for each fraction area, ϕ, are compiled in Figure 6a. Figure 6a shows the influence of the ratio between the channel area and the area of the plate (ϕ) on the once minimized filling time (t_m) for the I-shaped empty channel. Results in Figure 6a indicate that for the largest area fractions (ϕ) of the I-shaped channel, the filling time once minimized (t_m) increases. Figure 6b shows the influence of ϕ on the optimal ratio of L_0/H_0 named $(L_0/H_0)_o$. It is also noted that the behavior of the geometric ratio $(L_0/H_0)_o$ is modified with the variation of the area occupied by the channel, i.e., there is no universal optimal geometry that leads to the best performance for all area fractions of the open channels.

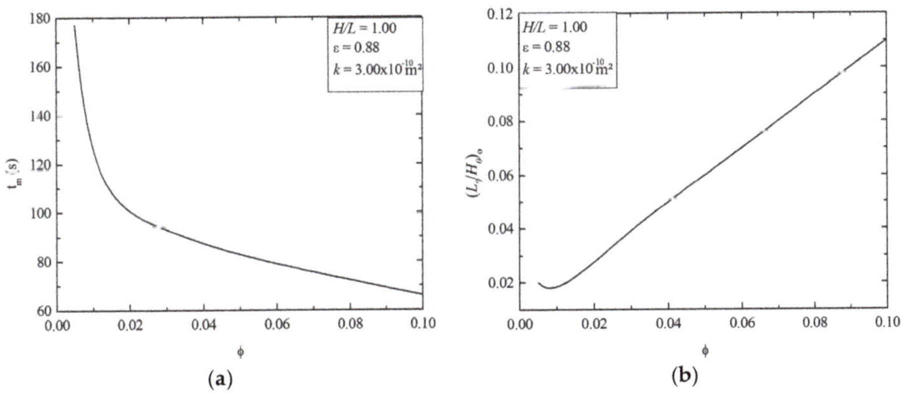

Figure 6. Effect of empty I-shaped channel area fraction (ϕ) over: (**a**) once minimized filling time (t_m), (**b**) once optimized L_0/H_0 ratio, $(L_0/H_0)_o$.

To illustrate the influence of ϕ and the L_0/H_0 ratio over resin flow considering the I-shaped channel, Figures 7 and 8 show the behavior of the resin advancement for three different instants of time at $\phi = 0.01$ and 0.1, respectively. The red region represents the resin, i.e., when the volume fraction is $f = 1.0$, while the blue region represents the air ($f = 0$). Regions with different colors represent resin/air mixture with intermediary volume fractions between resin and air ($0.0 < f < 1.0$). It is worth mentioning that this description is applied to all figures where the resin volume fraction is illustrated. Channel configuration is illustrated with the black line.

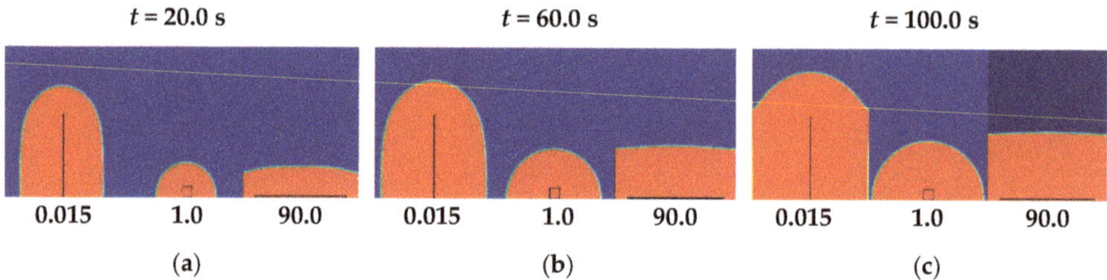

Figure 7. Resin volume fraction for ϕ = 0.01 and three different ratios of L_0/H_0: $(L_0/H_0)_o$ = 0.015, L_0/H_0 = 2.0, L_0/H_0 = 90.0 as a function of time: (**a**) t = 20.0 s, (**b**) t = 60.0 s, (**c**) t = 100 s.

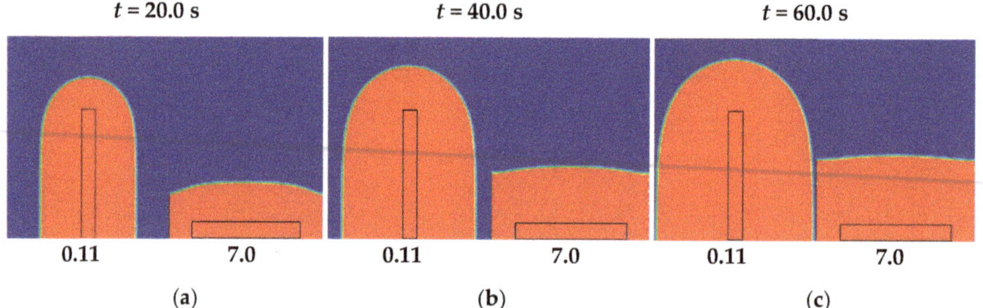

Figure 8. Resin volume fraction for ϕ = 0.1 and two different ratios of L_0/H_0: $(L_0/H_0)_o$ = 0.11 (optimal), L_0/H_0 = 7.0 (worst scenario) as a function of time: (**a**) t = 20.0 s, (**b**) t = 40.0 s, (**c**) t = 60.0 s.

Figure 7 shows resin advancement for ϕ = 0.010 and three different magnitudes of the ratio L_0/H_0: $(L_0/H_0)_o$ = 0.015, L_0/H_0 = 1.0 and L_0/H_0 = 90.0, which represents the two extreme magnitudes and an intermediate value of the ratio L_0/H_0, where the worst performance was reached. Three different instants of time are illustrated to show resin impregnation along time t = 20.0 s (Figure 7a); 60.0 s (Figure 7b) and 100.0 s (Figure 7c). Figure 8 shows the resin flows for ϕ = 0.100 and two different ratios of L_0/H_0: $(L_0/H_0)_o$ = 0.11 and L_0/H_0 = 7.0, representing the optimal geometry and worst case for three different instants of time: t = 20.0 s (Figure 8a), 40.0 s (Figure 8b), and 60.0 s (Figure 8c).

Results of Figures 7 and 8 indicated that for the smallest ratios of L_0/H_0, there is a higher advancement of resin in the y-direction. After the spread of resin along the entire empty channel, resin flows from the central region of the domain towards the lateral surfaces of the porous plate. Furthermore, in the superior region of the empty channel, the behavior of the resin flow is similar to that found in radial configurations. This behavior makes the distribution of the resin in the fibrous medium more efficient, minimizing resin filling time in the porous medium. It is also noticed that there are no void formations that could delay the infusion process or cause problems in the manufacturing process. For the highest magnitudes of L_0/H_0, the channel is placed in the lower region of the porous plate, leading to an almost constant front-line of resin.

Moreover, resin advances towards the exit of the porous domain with a behavior similar to the rectilinear case. In a general sense, geometrical investigation of I-shaped empty channels has attested that the elongated channels in resin propagation direction led to the best performance due to the increase of pressure gradient imposed inside the mold. This behavior is in agreement with the constructal principle of the optimal distribution of imperfections.

Subsequent investigation consisted of the evaluation of the influence of the T-shaped channel aspect ratios L_1/H_1, L_0/H_0, and H_0/H_1 on the filling time for resin impregnation in the rectangular porous plate. For this configuration, the optimization process was performed only for $\phi = 0.05$ due to the high number of simulations needed. Firstly, the effect of the ratio L_1/H_1 on resin infusion time along the mold was evaluated for different values of the ratio L_0/H_0, keeping H_0/H_1 fixed. Figure 9a–d present the effect of the ratio L_1/H_1 over the filling time for various constant ratios of L_0/H_0 and four different magnitudes of H_0/H_1 = 10.0, 20.0, 30.0 and 40.0, respectively. It should be noted that for H_0/H_1 = 30.0 and H_0/H_1 = 40.0, the L_0/H_0 search space becomes more restricted due to the generation of very thin channels. For example, when H_0/H_1 = 40.0, the variation is viable only in the range $0.05 \leq L_0/H_0 \leq 0.1$ because for higher magnitudes of L_1/H_1, the thickness of the bifurcated channel would become insignificant, which disfigures the T-shaped geometry of the channel and leads to the generation of permanent voids.

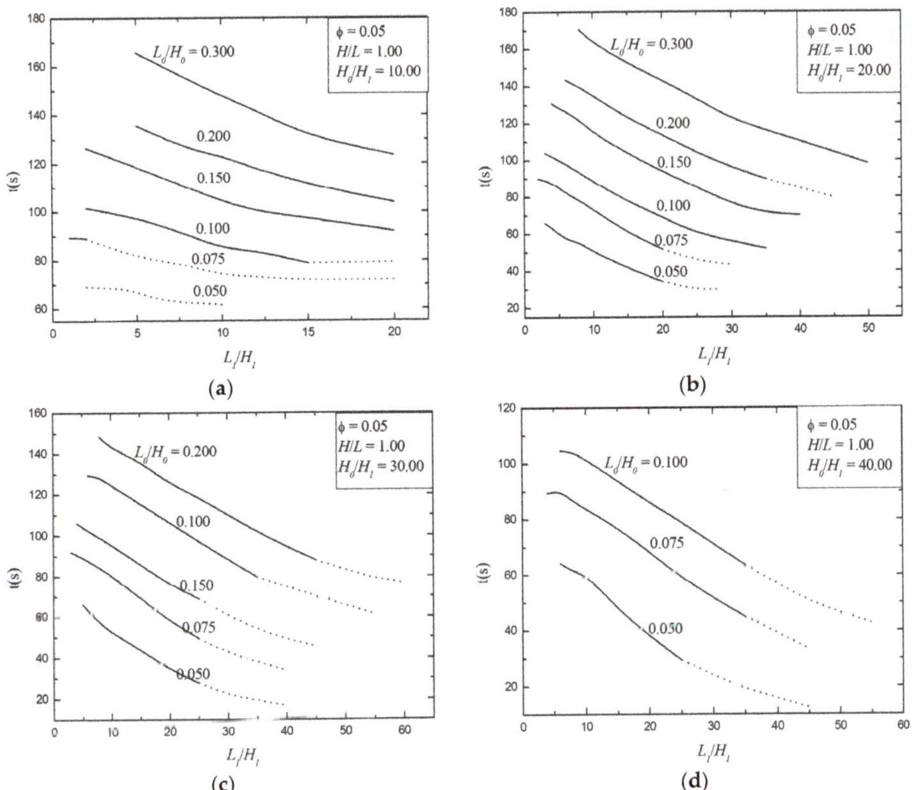

Figure 9. Effect of ratio L_1/H_1 over resin infusion filling time for several values of ratio L_0/H_0 and fixed values of H_0/H_1: (**a**) H_0/H_1 = 10.0, (**b**) H_0/H_1 = 20.0, (**c**) H_0/H_1 = 30.0, (**d**) H_0/H_1 = 40.0.

In Figure 9a–d, the continuous curves represent valid geometries, i.e., those geometries with no permanent void inside the porous plate or empty channel. The dashed curves represent cases where permanent voids are generated, even if the monitoring line indicates complete impregnation of the mold (with f = 1.0 along the whole measurement line). In this sense, these results are disregarded for optimization evaluation, since void formations are a critical problem in manufacturing. It is to be noted that some cases in the dashed region of the curves achieved lower filling times, indicating that the resin has a higher impregnation in the porous plate. However, the mold was not completely filled.

Results in Figure 9a–d also show that the higher magnitudes of L_1/H_1 for all investigated ratios of L_0/H_0 lead to the best performance, i.e., there is a reduction in resin filling time (t) when the bifurcated channel has higher penetration into the mold toward its lateral surfaces. However, some T-shaped channels with higher ratios of L_1/H_1 and lower magnitudes of L_0/H_0 lead to the formation of permanent voids, mainly when the geometry is composed of skinny channels. Therefore, the results recommend the building of elongated channels, since this does not lead to skinny channels. The generation of permanent voids occur, in general, in the corner that connects the single stem to the bifurcated branches. One possible reason for this behavior is the detachment of resin flow in the channel corner with not enough resin flow coming from the inlet to degenerate the void. Therefore, the permanent void remains during all the simulations in the area mentioned above.

To illustrate the influence of ratio L_1/H_1 on mold filling time, Figure 10 shows the advancement of resin for two different ratios of L_1/H_1, $L_1/H_1 = 6.0$ and $(L_1/H_1)_o = 35.0$, which represents the worst and best shapes when the constant ratios of $H_0/H_1 = 40.0$ and $L_0/H_0 = 0.1$ are assumed. For both cases, the volume fraction fields for three different instants of time—$t = 10.0$ s, 30.0 s, and 60.0 s—are presented in Figure 10a–c, respectively. For $L_1/H_1 = 6.0$, filling time of $t = 104.7$ s is obtained, while the ratio $(L_1/H_1)_o = 35.0$ leads to a resin injection time of the once minimized of $t_m = 63.5$ s, which is about 65.0% faster than the worst-performing geometry. For the lower magnitudes of H_1/L_1, resin distribution occurrs in a radial form in the upper region of the T-shaped channel and linearly from the center to the side surfaces of the mold at the bottom region of the T-shaped channel. This behavior is quite similar to that seen for I-shaped empty channels with high intrusion in the porous plate. As the ratio of H_1/L_1 is changed to $(H_1/L_1)_o = 35.0$, the radial advancement of the resin at the upper region of the T-shaped channel is found to be intensified, when compared with the previous case ($H_1/L_1 = 6.0$). This effect also intensifies the linear advancement of resin in the inferior region of the porous plate.

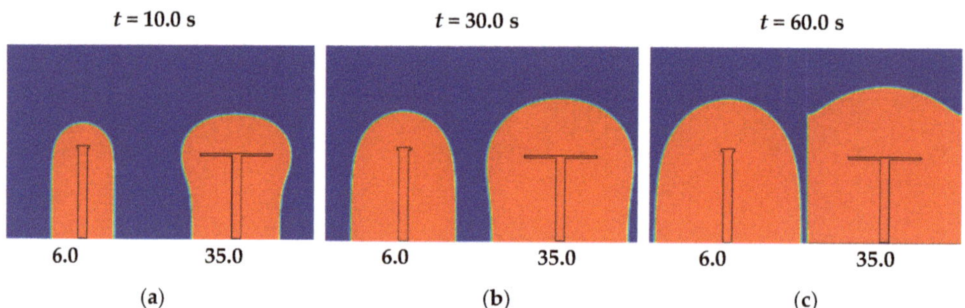

Figure 10. Resin volume fraction for $\phi = 0.05$, $H/L = 1.0$, $H_0/H_1 = 40.0$, $L_0/H_0 = 0.1$ and two different ratios of $L_1/H_1 = 6.0$ and $(L_1/H_1)_o = 35.0$ as a function of time: (**a**) $t = 10.0$ s, (**b**) $t = 30.0$ s, (**c**) $t = 60.0$ s.

Consequently, the filling of the resin in the horizontal direction (from the open channel towards the side walls) is completed in a shorter amount of time, also leading to an improvement in resin advancement towards the porous plate exit. However, significant increase in bifurcated length (augmentation of the ratio H_1/L_1) leads to the formation of permanent voids inside the plate. It is worth mentioning that the channel area is constant; as a consequence, augmentation of the channel bifurcated branch leads to a decrease in channel thickness, which is the main factor responsible for the formation of voids, especially in the corners between the single and bifurcated regions of the channel.

The variation of the L_1/H_1 ratio provides the once minimized filling time (t_m) of the resin in the porous plate for each investigated ratio of L_0/H_0 and different ratios of H_0/H_1. In the second optimization level, the results of Figure 9 are compiled and presented in Figure 11a,b.

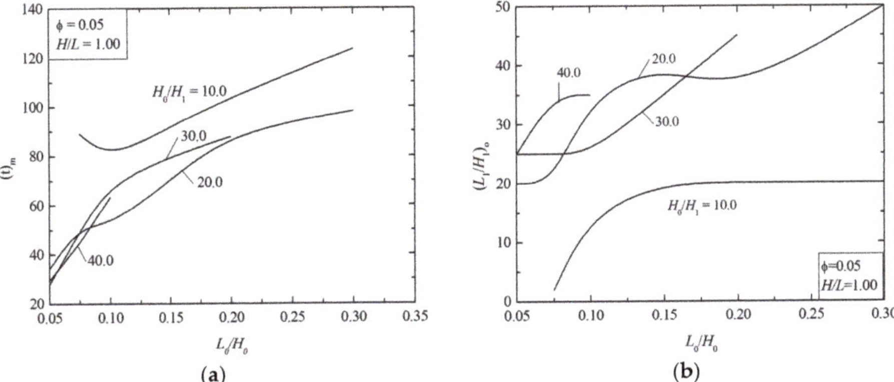

Figure 11. Effect of the ratio L_0/H_0 for different ratios of H_0/H_1 on (**a**) infusion time (once minimized), $(t)_m$; (**b**) ratio (once optimized), $(L_1/H_1)_o$.

Figure 11a illustrates the influence of the ratio L_0/H_0 over the once minimized filling time, t_m, for different H_0/H_1 ratios studied here. It can be noted that for H_0/H_1 = 20.0; 30.0 and 40.0, the effect of the L_0/H_0 over t_m is similar, except for cases with H_0/H_1 = 10.0. In general, it is noticed that lower ratios of L_0/H_0 present a predominance in achieving the twice-minimized filling time (t_{mm}), i.e., the best shapes are obtained when the main channel has a higher penetration in the direction of the exit region of the domain. For the ratio H_0/H_1 = 10.0, it can be observed that intermediate ratios of L_0/H_0 lead to the best performance of the problem. For the lowest magnitudes of L_0/H_0, the optimal ratios of $(L_1/H_1)_o$ suffer a strong decrease (see Figure 11b), degenerating the T-shaped channel into an I-shaped one. For higher magnitudes of L_0/H_0, the T-shaped channel is strongly restricted in the lower region of the porous plate. Figure 11b highlights the influence of the ratio L_0/H_0 over the once-optimized ratio of L_1/H_1, $(L_1/H_1)_o$, for the four different magnitudes of H_0/H_1 investigated. The effect of L_0/H_0 over $(L_1/H_1)_o$ varied for different magnitudes of H_0/H_1. For the ratios H_0/H_1 = 20.0; 30.0 and 40.0, the best shapes are obtained for $(L_0/H_0)_o$ = 0.05 and for high magnitudes of $(L_1/H_1)_{oo}$ in the range $20.0 \leq (L_1/H_1)_{oo} \leq 25.0$, corroborating previous findings that the best performance is achieved for the most elongated possible single and bifurcated branches of T-shaped channel. For H_0/H_1 = 10.0, the best configurations changed for $(L_0/H_0)_o$ = 0.075 and the optimal ratio $(L_1/H_1)_{oo}$ dropped dramatically to $(L_1/H_1)_{oo}$ = 2.0. In general, the results show that changes in one degree of freedom have a strong influence on the effect of other geometric ratios on the performance indicator. In this case, it was also noticed that there is no optimal universal configuration that leads to the best performance in this problem.

The last optimization step consists of investigation on the influence of the ratio H_0/H_1 over the twice-minimized filling time (t_{mm}) and the respective optimal configurations: $(L_0/H_0)_o$ and $(L_1/H_1)_{oo}$. To obtain this evaluation, the best results achieved in Figure 11a,b are compiled in Figure 12.

These results show that the behavior of the (t_{mm}) is largely affected by ratio H_0/H_1, and it has a strong dependence on ratios $(L_1/H_1)_{oo}$ and $(L_0/H_0)_o$. The thrice-minimized filling time (t_{mmm}) is obtained for the rate $(H_0/H_1)_o$ = 30.0, which is only 5.0% lower than that reached for H_0/H_1 = 40.0. In general, it is possible to state that the higher magnitudes of the ratio H_0/H_1 led to the best performance for this problem. The lowest magnitude of H_0/H_1 = 10.0 obtained a performance 183.0% inferior to that reached for the optimal ratio $(H_0/H_1)_o$ = 30.0. Moreover, differences found for t_{mm} between the ratios H_0/H_1 = 20.0 and $(H_0/H_1)_o$ = 30.0 was nearly 23.0%.

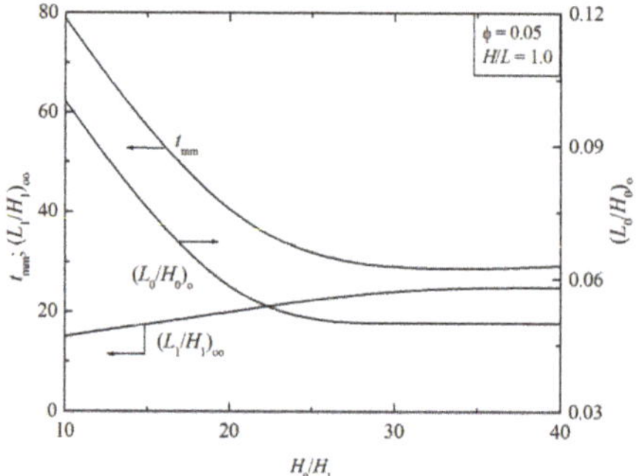

Figure 12. Effect of the ratio H_0/H_1 for twice-minimized infusion time, t_{mm}, and the respective optimal shapes, $(L_0/H_0)_o$, $(L_1/H_1)_{oo}$.

Concerning optimal shapes, it was noticed that the ratio $(L_0/H_0)_o$ suffered a strong decrease in its magnitude in the region $10.0 \leq H_0/H_1 \leq 20.0$, followed by a smooth decrease in the range $20.0 \leq H_0/H_1 \leq 30.0$ and a stabilization for $H_0/H_1 \geq 30.0$. For $(L_1/H_1)_{oo}$, an increase in magnitude was noticed, until stabilization for $H_0/H_1 \geq 30.0$. In general, the results show the importance of geometric investigation in the reduction of filling time of resin impregnation along a porous domain.

To illustrate the transient behavior of resin flow in the mold, Figures 13 and 14 show the distribution of resin volume fraction as a function of time for the worst performing ratio, $H_0/H_1 = 10.0$, and for the optimal ratio $(H_0/H_1)_o = 30.0$, respectively. Topologies are presented for the following times: $t = 20.0$ s; 40.0 s and 60.0 s, respectively. Figure 7, Figure 8, and Figure 10 illustrate the topology of the configuration of the empty channel.

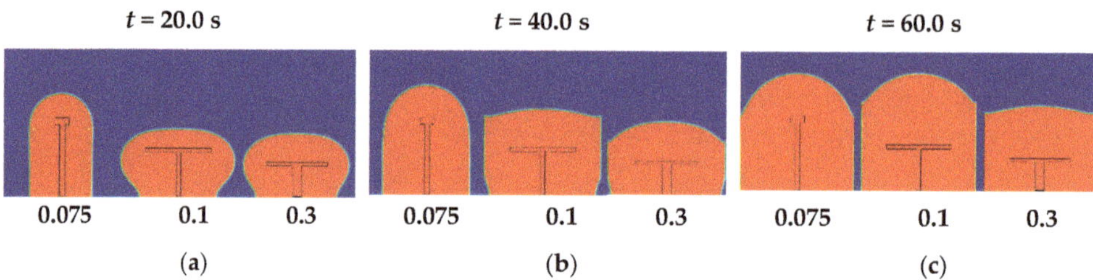

Figure 13. Resin volume fraction for $\phi = 0.05$, $H/L = 1.0$, $H_0/H_1 = 10.0$, for three different ratios of $L_0/H_0 = 0.075$, $(L_0/H_0)_o = 0.100$ and $L_0/H_0 = 0.300$ as a function of time: (**a**) $t = 20.0$ s, (**b**) $t = 40.0$ s, (**c**) $t = 60.0$ s.

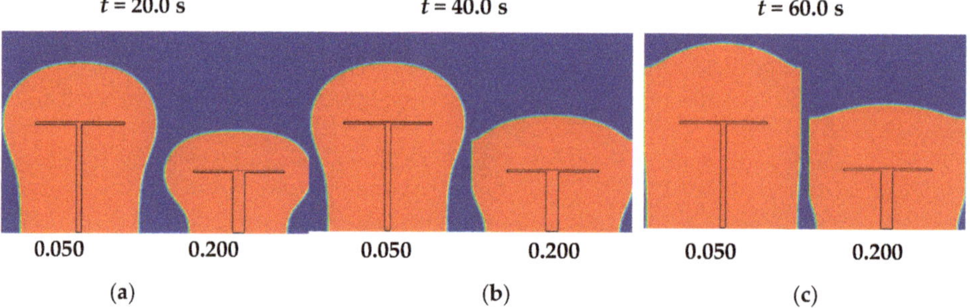

Figure 14. Resin volumetric fraction for $\phi = 0.05$, $H/L = 1.0$, $H_0/H_1 = 30.0$, for two different ratios of $(L_0/H_0)_{oo} = 0.050$ and $L_0/H_0 = 0.200$ as a function of time: (**a**) $t = 20.0$ s, (**b**) $t = 40.0$ s, (**c**) $t = 60.0$ s.

Figure 13 shows resin progression at the analyzed instants of time for $H_0/H_1 = 10.0$ and $L_0/H_0 = 0.075$, $(L_0/H_0)_o = 0.1$, and $L_0/H_0 = 0.3$ corresponds to the lowest ratio of L_0/H_0, the optimal configuration, and the superior magnitude, respectively. For the lowest ratio of L_0/H_0, it is observed that the T-shaped channel almost degenerates into an I-shaped channel. This configuration leads to a poor distribution of resin in the lower region of the porous plate. The resin propagation is shown for three different instants of time: $t = 20.0$ s (Figure 13a), $t = 40.0$ s (Figure 13b) and $t = 60.0$ s (Figure 13c).

For optimal configuration, despite the T-shaped channel being placed in a lower region of the porous plate in comparison with the case with $L_0/H_0 = 0.075$, the elongated bifurcated branches led to better distribution of the resin in the lower region of the porous plate and allow easy resin propagation towards the exit of the domain. For the superior ratio of $L_0/H_0 = 0.3$, the restriction of the T-shaped channel in the lower region of the porous plate affects the advancement of resin towards the exit of the domain, leading to a poorer performance than the optimal ratio of L_0/H_0.

Figure 14 presents the results of resin spread for the optimum ratio of H_0/H_1, $(H_0/H_1)_o = 30.0$, for two different ratios of L_0/H_0, $(L_0/H_0)_{oo} = 0.05$, $L_0/H_0 = 0.2$ and three different instants of time: $t = 20.0$ s (Figure 14a), $t = 40.0$ s (Figure 14b), $t = 60.0$ s (Figure 14c). These ratios represent the corresponding twice-optimized and worst-case scenarios for the ratio of $(H_0/H_1)_o = 30.0$, respectively. For the ratio $(L_0/H_0)_{oo} = 0.05$, an injection time of $t_{mmm} = 27.9$ s is obtained, and the ratio $L_0/H_0 = 0.2$ leads to a once minimized injection time of $t_m = 87.9$ s, which is, about of 215.0% higher than the previously obtained optimal geometry. It is then possible to significantly improve the performance of the studied system even in the last level of optimization.

In a general sense, the results demonstrate that for the T-shaped channel, all geometrical ratios (degrees of freedom) affect the filling time of resin impregnation in a porous domain. They also show that one degree of freedom has a strong influence over the optimal ratios of other geometric ratios, e.g., the ratio H_0/H_1 affected the optimal ratios of $(L_1/H_1)_{oo}$ and $(L_0/H_0)_o$.

All the geometrical parameters studied here (L_1/H_1, L_0/H_0, and H_0/H_1) has some sensibility in the infusion process and cannot be neglected. Moreover, a comparison between the performance of the T-shaped and I-shaped empty channels, for the same conditions, demonstrates that the thrice-optimized T-shaped channel leads to performance that is almost three times superior to that achieved by the optimal I-shaped configuration.

In the present fluid dynamic conditions, where resin flow has little advective dominance, the rectangular porous plate can be viewed as an elementary construction of multiple structures generated by the replication of a porous plate in x- or y-axes. In this case, multiple optimal configurations of I and T-shaped channels can be associated in a series and/or parallel using the optimal configurations reached in this work.

5. Conclusions

This paper developed a numerical study of the geometric evaluation of empty channels inserted in a porous rectangular plate, mimicking the liquid resin infusion process (LRI), and applying constructal design and exhaustive search (ES) for geometric optimization. The primary purposes here were to analyze the influence of the design of I and T-shaped empty channels on the filling time of resin impregnation in a porous plate, preventing the generation of permanent voids.

The following important recommendations were made:

- I-shaped channels with the highest penetration in the y-direction of the porous domain, as well as the highest magnitude of area fraction (ϕ) had a higher performance in minimization of filling time, as expected.
- For the T-shaped channels, the results demonstrated that the best performance was reached for the highest possible penetration of single and bifurcated channels in the porous plate, with the exception of cases with very thin channels that saw the generation of permanent voids.

- The results for the T-shaped channels demonstrated that all studied geometric ratios had sensibility over the performance indicator (filling time) and the optimal ratios. For example, the ratio H_0/H_1 had a strong influence on the twice-minimized filling time (t_{mm}) and the respective optimal ratios, $(L_1/H_1)_{oo}$ and $(L_0/H_0)_o$.
- For the same fluid dynamic conditions, the results indicated that the most complex empty T-shaped channel was beneficial in improving resin flow impregnation in the whole domain of the porous plate, as compared to the elementary I-shaped one.
- In general, the results of this study demonstrated the importance of constructal design in the geometric evaluation of the liquid resin infusion process.

Other geometrical configurations, such as Y-shaped and fishbone empty channels, or geometries being constructed with no pre-defined form are recommended for future studies. Another interesting possibility of study is the insertion of two or more inlet ports investigating the influence of interactions between different channels on their optimal shapes.

Author Contributions: Conceptualization, J.A.S., and E.D.d.S.; methodology, G.M.C.M., L.A.I., and R.d.L.L.; software, G.M.C.M., J.A.S., and E.D.d.S.; validation, L.A.I., C.F., G.M.C.M., and J.A.S.; formal analysis, E.D.d.S., L.A.O.R., J.A.S., and C.F.; investigation, G.M.C.M., E.D.d.S., and J.A.S.; resources, S.C.A., J.A.S., and E.D.d.S.; data curation, E.D.d.S., and J.A.S.; writing—original draft preparation, G.M.C.M., R.d.L.L., and E.D.d.S.; writing—review and editing, E.D.d.S., L.A.I., and C.F.; visualization, J.S. and L.A.O.R.; supervision, G.M.C.M., L.A.O.R., and C.F.; project administration, E.D.d.S. and J.A.S.; funding acquisition, G.M.C.M., J.A.S., and E.D.d.S. All authors have read and agreed to the published version of the manuscript.

Funding: This research was funded by the Brazilian Coordination for the Improvement of Higher Education Personnel (CAPES) (Finance Code 001), the Brazilian National Council for Scientific and Technological Development (CNPq), and the Italian Minister of Foreign Affairs and International Cooperation (MAECI), as part of the 'Two Seats for a Solar Car' international project.

Acknowledgments: The authors thank Giangiacomo Minak for supporting this publication.

Conflicts of Interest: The authors declare no conflict of interest. The funders had no role in the design of the study; in the collection, analyses, or interpretation of data; in the writing of the manuscript, or in the decision to publish the results.

Nomenclature

A	Area of the rectangular porous medium, m^2
A_I	Area of the I-shaped empty channel, m^2
A_T	Area of the T-shaped empty channel, m^2
$\vec{F}$	External forces per unit volume, N/m^3
f	Volume fraction, [-]
H	Height of rectangular porous mold, m
H_0	Length of a single channel, m
H_1	Thickness of the bifurcated channel, m
K	Permeability of the medium, m^2
L	Length of rectangular porous mold, m
L_0	Thickness of the single-channel, m
L_1	Length of the bifurcated channel, m
P	Total pressure, Pa
P_{in}	Injection pressure, Pa
P_{out}	Exit pressure, Pa
t	Time, s
$\vec{V}$	Velocity vector, m/s
x, y	Spatial coordinates of a two-dimensional domain, m
X_f	Position of the advancement of resin front line in verification case, m

Greek symbols

ρ	Density, kg/m^3
μ	Fluid dynamic viscosity, Pa s
ε	Porosity, [-]
ϕ	The ratio between the empty channel and the porous plate areas, [-]
$\bar{\bar{\tau}}$	Stress tensor, N/m^2

Subscripts

o	Once optimized
oo	Twice optimized
ooo	Thrice optimized
m	Once minimized
mm	Twice minimized
mmm	Thrice minimized
T	Transposed
0	Single-channel
1	Bifurcated channel

References

1. Wang, P.; Drapier, S.; Molimard, J.; Vautrin, A.; Minni, J. Numerical and experimental analyses of resin infusion manufacturing processes of composite materials. *J. Compos. Mater.* **2012**, *46*, 1617–1631. [CrossRef]
2. Yang, J.; Jia, Y.; Ding, Y.; He, H.; Shi, T.; An, L. Edge effect in RTM processes under constant pressure injection conditions. *J. Appl. Polym. Sci.* **2010**, *118*, 1014–1019. [CrossRef]
3. Brouwer, W.D.; van Herpt, E.C.F.C.; Labordus, M. Vacuum injection moulding for large structural applications. *Compos. Part A Appl. Sci. Manuf.* **2003**, *34*, 551–558. [CrossRef]
4. Goncharova, G.; Cosson, B.; Lagardère, M.D. Analytical modeling of composite manufacturing by vacuum assisted infusion with minimal experimental characterization of random fabrics. *J. Mater. Process. Technol.* **2015**, *219*, 173–180. [CrossRef]
5. Pierce, R.S.; Falzon, B.G. Simulating resin Infusion through textile reinforcement materials for the manufacture of complex composite structures. *Engineering* **2017**, *3*, 596–607. [CrossRef]
6. Rondina, F.; Taddia, S.; Mazzocchetti, L.; Donati, L.; Minak, G.; Rosenberg, P.; Bedeschi, A.; Dolcini, E. Development of full carbon wheels for sport cars with high-volume technology. *Compos. Struct.* **2018**, *192*, 368–378. [CrossRef]
7. Poodts, E.; Minak, G.; Mazzocchetti, L.; Giorgini, L. Fabrication, process simulation and testing of a thick CFRP component using the RTM process. *Compos. Part B Eng.* **2014**, *56*, 673–680. [CrossRef]
8. Poodts, E.; Minak, G.; Dolcini, E.; Donati, L. FE analysis and production experience of a sandwich structure component manufactured by means of vacuum assisted resin infusion process. *Compos. Part B Eng.* **2013**, *53*, 179–186. [CrossRef]
9. Isoldi, L.A.; Oliveira, C.P.; Rocha, L.A.O.; Souza, J.A.; Amico, S.C. Three-dimensional numerical modeling of RTM and LRTM processes. *J. Braz. Soc. Mech. Sci. Eng.* **2012**, *34*, 105–111. [CrossRef]
10. Grössing, H.; Stadlmajer, N.; Fauster, E.; Fleischmann, M.; Schledjewski, R. Flow front advancement during composite processing: Predictions from numerical filling simulation tools in comparison with real-world experiments. *Polym. Compos.* **2016**, *37*, 2782–2793. [CrossRef]
11. Sirtautas, J.; Pickett, A.K.; George, A. Materials characterization and analysis for flow simulation of liquid resin infusion. *Appl. Compos. Mater.* **2015**, *22*, 323–341. [CrossRef]
12. Pierce, R.S.; Falzon, B.G.; Thompson, M.C. A multi-physics process model for simulating the manufacture of resin-infused composite aerostructures. *Compos. Sci. Technol.* **2017**, *149*, 269–279. [CrossRef]
13. Chebil, N.; Deléglise-Lagardère, M.; Park, C.H. Efficient numerical simulation method for three dimensional resin flow in laminated preform during liquid composite molding processes. *Compos. Part A Appl. Sci. Manuf.* **2019**, *125*, 105519. [CrossRef]
14. Rubino, F.; Carlone, P. A semi-analytical model to predict infusion time and reinforced thickness in VARTM and SCRIMP processes. *Polymers* **2019**, *11*, 20. [CrossRef]
15. Falaschetti, M.P.; Rondina, F.; Zavatta, N.; Gragnani, L.; Gironi, M.; Troiani, M.; Donati, L. Material Characterization for Reliable Resin Transfer Molding Process Simulation. *Appl. Sci.* **2020**, *10*, 1814. [CrossRef]

16. Mathur, R.; Advani, S.G.; Fink, B.K. Use of genetic algorithms to optimize gate and vent locations for the resin transfer molding process. *Polym. Compos.* **1999**, *20*, 167–178. [CrossRef]
17. Jiang, S.; Zhang, C.; Wang, B. A process performance index and its application to optimization of the RTM process. *Polym. Compos.* **2001**, *22*, 690–701. [CrossRef]
18. Luo, J.; Liang, Z.; Zhang, C.; Wang, B. Optimum tooling design for resin transfer molding with virtual manufacturing and artificial intelligence. *Compos. Part A Appl. Sci. Manuf.* **2001**, *32*, 877–888. [CrossRef]
19. Ratle, F.; Achim, V.; Trochu, F. Evolutionary operators for optimal gate location in liquid composite moulding. *Appl. Soft Comput.* **2009**, *9*, 817–823. [CrossRef]
20. Wang, J.; Simacek, P.; Advani, S.G. Use of centroidal Voronoi diagram to find optimal gate locations to minimize mold filling time in resin transfer molding. *Compos. Part A Appl. Sci. Manuf.* **2016**, *87*, 243–255. [CrossRef]
21. Gomes, P.P.; Ferro, O.A.G.; Rezende, M.C. Experimental characterization and simulation of VARTM process to obtain carbon/epoxi composites. In Proceedings of the 1st Brazilian Conference on Composite Materials (BCCM1), Natal, Brazil, 16–19 July 2012.
22. Struzziero, G.; Skordos, A.A. Multi-objective optimization of resin infusion. *Adv. Manuf. Polym. Compos. Sci.* **2019**, *5*, 17–28. [CrossRef]
23. Geng, Y.; Jiang, J.; Chen, N. Local impregnation behavior and simulation of non-crimp fabric on curved plates in vacuum assisted resin transfer molding. *Compos. Struct.* **2019**, *208*, 517–524. [CrossRef]
24. Shevtsov, S.; Zhilyaev, I.; Chang, S.-H.; Wu, J.-K.; Huang, J.-P.; Snezhina, N. Experimental and numerical study of vacuum resin infusion for thin-walled composite parts. *Appl. Sci.* **2020**, *10*, 1485. [CrossRef]
25. Bejan, A. *Shape and Structure, from Engineering to Nature*; Cambridge University Press: New York, NY, USA, 2000.
26. Bejan, A.; Lorente, S. *Design with Constructal Theory*; John Wiley & Sons: Hoboken, NJ, USA, 2008.
27. Bejan, A.; Zane, P. *Design in Nature, How the Constructal Law Governs Evolution in Biology, Physics, Technology, and Social Organization*, 1st ed.; Doubleday: New York, NY, USA, 2012.
28. Bejan, A. *The Physics of Life, the Evolution of Everything*; St. Martins Press: New York City, NY, USA, 2016.
29. Bejan, A. Thermodynamics today. *Energy* **2018**, *160*, 1208–1219. [CrossRef]
30. Xie, G.; Song, Y.; Asadi, M.; Lorenzini, G. Optimization of Pin-Fins for a Heat Exchanger by Entropy Generation Minimization and Constructal Law. *J. Heat Transf.* **2015**, *137*, 061901. [CrossRef]
31. Hajmohammadi, M.R. Introducing a ψ-shaped cavity for cooling a heat generating medium. *Int. J. Therm. Sci.* **2017**, *121*, 204–212. [CrossRef]
32. Mirzaei, M.; Hajabdollahi, H.; Fadakar, H. Multi-objective optimization of shell-and-tube heat exchanger by constructal theory. *Appl. Therm. Eng.* **2017**, *125*, 9–19. [CrossRef]
33. Bello-Ochende, T.; Olakoyejo, O.T.; Meyer, J.P.; Bejan, A.; Lorente, S. Constructal flow orientation in conjugate cooling channels with internal heat generation. *Int. J. Heat Mass Transf.* **2013**, *57*, 241–249. [CrossRef]
34. Martins, J.C.; Goulart, M.M.; Gomes, M.N.; Souza, J.A.; Rocha, L.A.O.; Isoldi, L.A.; dos Santos, E.D. Geometric evaluation of the main operational principle of an overtopping wave energy converter by means of Constructal Design. *Renew. Energy* **2018**, *118*, 727–741. [CrossRef]
35. Feng, H.; Chen, L.; Xie, Z. Multi-disciplinary, multi-objective and multi-scale constructal optimizations for heat and mass transfer processes performed in Naval University of Engineering, a review. *Int. J. Heat Mass Transf.* **2017**, *115*, 86–98. [CrossRef]
36. Bejan, A.; Gunes, U.; Sahin, B. The evolution of air and maritime transport. *Appl. Phys. Rev.* **2019**, *6*, 021319. [CrossRef]
37. Isoldi, L.A.; Souza, J.A.; dos Santos, E.D.; Marchesini, R.; Porto, J.; Letzow, M.; Rocha, L.A.O.; Amico, S.C. Constructal design applied to the light resin transfer molding (LRTM) manufacturing process. In Proceedings of the 22nd ABCM International Congress of Mechanical Engineering (COBEM), Ribeirao Preto, Brazil, 3–7 November 2013.
38. Magalhães, G.M.C.; Lorenzini, G.; Nardi, M.G.; Amico, S.C.; Isoldi, L.A.; Rocha, L.A.O.; Souza, J.A.; dos Santos, E.D. Geometrical evaluation of a resin infusion process by means of constructal design. *Int. J. Heat Technol.* **2016**, *34*, S101–S108. [CrossRef]
39. Patankar, S.V. *Numerical Heat Transfer and Fluid Flow*; McGraw Hill: New York, NY, USA, 1980.
40. Versteeg, H.K.; Malalasekera, W. *An Introduction to Computational Fluid Dynamics, the Finite Volume Method*; Pearson: London, UK, 2007.
41. FLUENT Inc. *Documentation Manual—FLUENT 14.3*; FLUENT Inc.: New York, NY, USA, 2013.

42. Hirt, C.W.; Nichols, B.D. Volume of fluid (VOF) method for the dynamics of free boundaries. *J. Comput. Phys.* **1981**, *39*, 201–225. [CrossRef]
43. Schlichting, H. *Boundary Layer Theory*, 7th ed.; McGraw-Hill: New York, NY, USA, 1979.
44. Morren, G.; Bottiglieri, M.; Bossuyt, S.; Sol, H.; Lecompte, D.; Verleye, B.; Lomov, S.V. A reference specimen for permeability measurements of fibrous reinforcements for RTM. *Compos. Part A Appl. Sci. Manuf.* **2009**, *40*, 244–250. [CrossRef]
45. Rudd, C.D. *Liquid Moulding Technologies, Resin Transfer Moulding, Structural Reaction Injection Moulding and Related Processing Techniques*; Woodhead Publ: Cambridge, UK, 1997.
46. Srinivasan, V.; Salazar, A.J.; Saito, K. Modeling the disintegration of modulated liquid jets using volume-of-fluid (VOF) methodology. *Appl. Math. Model.* **2011**, *35*, 3710–3730. [CrossRef]
47. Dos Santos, E.D.; Isoldi, L.A.; Gomes, M.N.; Rocha, L.A.O. *The Constructal Design Applied to Renewable Energy Systems*; Rincón-Mejía, E., de las Heras, A., Eds.; CRC PRESS: Boca Raton, FL, USA, 2017.
48. Gonzales, G.V.; Estrada, E.S.D.; Emmendorfer, L.R.; Isoldi, L.A.; Xie, G.; Rocha, L.A.O.; Dos Santos, E.D. A comparison of simulated annealing schedules for constructal design of complex cavities intruded into conductive walls with internal heat generation. *Energy* **2015**, *93*, 372–382. [CrossRef]
49. Radiša, R.; Dučić, N.; Manasijević, S.; Marković, N.; Ćojbašić, Ž. Casting improvement based on metaheuristic optimization and numerical simulation. *Facta Univ.-Ser. Mech. Eng.* **2017**, *15*, 397–411. [CrossRef]
50. Hu, J.; Liu, Y.; Shao, X. Study on void formation in multi-layer woven fabrics. *Compos. Part A Appl. Sci. Manuf.* **2004**, *35*, 595–603. [CrossRef]
51. Schmidt, T.M.; Goss, T.M.; Amico, S.C.; Lekakou, C. Permeability of hybrid reinforcements and mechanical properties of their composites molded by resin transfer molding. *J. Reinf. Plast. Compos.* **2009**, *28*, 2839–2850. [CrossRef]

Article

Material Characterization for Reliable Resin Transfer Molding Process Simulation

Maria Pia Falaschetti [1,*], Francesco Rondina [1], Nicola Zavatta [1], Lisa Gragnani [2], Martina Gironi [1], Enrico Troiani [1,*] and Lorenzo Donati [1]

1 University of Bologna, Department of Industrial Engineering, Via Fontanelle 40, 47121 Forlì (FC), Italy; francesco.rondina2@unibo.it (F.R.); nicola.zavatta2@unibo.it (N.Z.); l.donati@unibo.it (L.D.); martinagironi@gmail.com (M.G.)

2 ESI Italia, Viale Angelo Masini, 36, 40126 Bologna (BO), Italy; lisa.gragnani@esi-group.com

* Correspondence: mariapi.falaschetti2@unibo.it (M.P.F.); enrico.troiani@unibo.it (E.T.); Tel.: +39-0543-374449 (M.P.F. & E.T.)

Received: 11 February 2020; Accepted: 3 March 2020; Published: 6 March 2020

Abstract: Resin transfer molding (RTM) technologies are widely used in automotive, marine, and aerospace applications. The need to evaluate the impact of design and production critical choices, also in terms of final costs, leads to the wider use of numerical simulation in the preliminary phase of component development. The main issue for accurate RTM analysis is the reliable characterization of the involved materials. The aim of this paper is to present a validated methodology for material characterization to be implemented and introduce data elaboration in the ESI PAM-RTM software. Experimental campaigns for reinforcement permeabilities and resin viscosity measurement are presented and discussed. Finally, the obtained data are implemented in the software and then compared to experimental results in order to validate the described methodology.

Keywords: RTM; composites; FEM simulation; permeability characterization

1. Introduction

Fiber-reinforced composite materials are widely used in several lightweight applications in the nautical, aerospace, and automotive sectors due to high strength-to-density ratios. More extensive applications are limited by materials and technology costs. Even if the cost of carbon and glass fibers follows a decreasing trend, the manufacturing costs will still be almost constant if conventional high-performance technologies are used (i.e., autoclave manufacturing). Moreover, several composite technologies cannot provide a component cost reduction by increasing the production volume. To achieve a reliable component cost reduction, new out-of-autoclave technologies (like infusion technologies, prepreg compression technologies, or short-fiber compound processing) have been recently optimized for industrial use.

Among the others, resin transfer molding (also known as light RTM), and its variants vacuum-assisted resin infusion (VARI) and high-pressure resin transfer molding (HP-RTM), are promising technologies for producing complex components with high performance at a higher volume rate. Moreover, these processes are economically advantageous with respect to the others, such as autoclave curing of pre-impregnated layers, due to cheaper materials and shorter processing times.

RTM technologies consist of placing a dry three-dimensional preform into a mold cavity and injecting resin-hardener mixture into the closed mold. The injection pressure (assisted by a vacuum for VARI or high pressure in HP-RTM processing), together with fiber permeability, is able to produce components with 50% and higher fiber volume fraction, thus providing excellent mechanical properties [1–3]. Nevertheless, in the light RTM process, the low flow rate of resin and low injection pressure result in a long resin injection time, thus hampering the use of fast-cure resins and, consequently, the overall productivity. As a consequence, the light RTM process is mainly limited to the low-volume manufacturing capability and, only to a minor extent, to slightly lower mechanical properties related to the lower fiber volume fraction. High-volume manufacturing with RTM is possible only if the process cycle time is significantly reduced. VARI and HP-RTM variants are improvements in this direction. Due to the differential pressure between inlet and outlet flows (VARI) or to higher flow rates (HP-RTM), together with the choice of adequate resin and hardener, production time can be reduced. A critical point of RTM technologies is the assurance of a full impregnation of the component, which is influenced by reinforcement types, the number of layers, resin viscosity, position of resin injection (ports), and vacuum extraction (vents). As molds can reach very high manufacturing costs when wide components are involved or when metal molds are required, tools supporting the design and critical process selection are needed in order to reduce uncertainties already in the preliminary phase of component development.

In this direction, Finite Element Method (FEM) software has been developed in the last few years to assist in component design and processing, predicting issues such as resin-rich areas, air bubbles, dry spots, zones of high porosity, etc., and for optimization of mold geometry and process parameters, such as port and vent locations [4–7]. In the case of critical molds, therefore, it is advantageous to perform a preliminary study by means of numerical simulation in order to avoid expensive experimental setups and prototypes.

In this context, the ESI group developed a PAM-RTM module, which can solve liquid resin infusion processes, like light and high-pressure RTM, compression resin transfer molding (C-RTM), and VARI. ESI's approach, resulting from several years of collaboration with academics [8–10], consists of a finite element solver based on the coupling of resin flow (governed by Darcy's law) and the preform behavior (considered as a porous medium undergoing deformations). Thus, the solver can provide the filling time and properties (thicknesses, fiber volume contents, geometry) of the final product [11].

However, accurate information on the material properties is needed for reliable simulations of the RTM process: in particular, reinforcement permeabilities and resin viscosity. Several works in the literature are available for the experimental determination of the permeability of RTM reinforcements [12]. Most of them evaluate the in-plane permeability [13,14], concluding that radial flow (liquid injected from transverse direction at well-defined locations to evaluate 2D permeability) and linear flow (liquid injected from one end of the composite preform to evaluate 1D permeability) experiments give consistent results. The through-the-thickness permeability is more complex to determine [15–21] and is usually neglected because of the small thicknesses of the preforms.

Therefore, a well-defined procedure for the experimental determination of the permeability tensor of fiber textiles and resin viscosity has been formulated in this paper in order to provide a reliable method to simulate RTM production of industrial components using PAM-RTM software. The key steps, from material characterization to numerical simulations, are illustrated by referring to the production of a real component in order to compare to a real-life test case. The permeabilities of four different types of textile reinforcements (three carbon fibers and one glass fiber) were experimentally measured. Moreover, the viscosity of a synthetic oil (15W40, used in glass textile permeability tests) and orthophthalic resin (Lavesan LERPOL 666/S RAL 9010) was tested. Permeability tests were modelled in the case of carbon fiber reinforcement tests, while in the case of glass fibers (GFRP), a naval component was reproduced. The analyzed GFRP part represents coverage for an aft-peak, placed aft-ward the cockpit, and it belongs to a series of pieces that are mounted on a 15.85 m (52 ft) cruising sailing yacht. This choice was sustained by the wider experience and higher amount of available data that RTM technology shows in marine industry compared to automotive, where production information represents more sensible data.

2. Materials and Methods

2.1. Test Material

The naval component was made of a polyester resin reinforced by glass fibers. In particular, the following materials were used:

- a stitched glass fiber textile, i.e., SAERTEX SAERcore Max BX600/PP18/450, made of a layer of 600 g/m^2 E-glass biaxial fibers [+/−45° orientation] and a 450 g/m^2 chopped strand mat stitched by means of a 18 g/m^2 PP (polypropylene flow media layer). The nominal thickness of the dry textile was 3.2 mm.
- orthophthalic resin Lavesan LERPOL 666/S RAL 9010. The resin was diluted with 2% of styrene and added with 1% of catalyst (CUROX M-303 methyl ethyl ketone peroxide) prior to injection into the mold.
- an ultralight foam made of polyurethane, polyethylene, and polyisocyanurate, i.e., SAERTEX Saerfoam. Good resin flow through the core thickness was achieved by 1 mm diameter holes within the foam, spaced 20 mm from one another. The holes were filled with glass fibers to increase shear and compression resistance. The nominal thickness of the foam was 10 mm.

The nominal properties of the resin are reported in Table 1. Accurate characterization of the flow properties of the resin is vital for manufacturing control, as well as for numerical modelling of the process. To this end, experimental tests were conducted to assess the resin viscosity and the fiber volume fraction and permeability of the textile.

Table 1. Resin properties.

Propriety	Testing Standard	Value
Absolute density at 25 °C	ISO 1675/75	1.06 g/mL
Brookfield Viscosity at 25 °C	MA 041 LCP	300 ± 50 cPs
Gel time at 25 °C	MA 120-LR	6′00″ ± 1′30″
Exothermic peak	MA 120-LR	178 ± 2 °C

Moreover, in order to provide references respect to other fibres and styles, carbon fiber reinforcement textiles were tested in permeability tests. These materials are a unidirectional fiber 12K T700 300 g/m^2, a triaxial fiber (30/90/−30) 24K T700 300 g/m^2, and a Twill 2×2 fabric 12K T700 630 g/m^2.

2.2. Methods Used for Material Properties Characterization

Resin viscosity was measured by using a Thermo Scientific Brookfield Haake 7 plus viscosimeter. Four values of the rotational speed were used for testing, i.e., 20, 30, 50, and 100 rpm, considering both room and production temperature. The viscosity of the oil employed for the permeability tests was also measured by using the same apparatus.

The textile permeability was assessed by means of unsaturated linear flow tests. Assuming that the textile reinforcement in an RTM component is a porous medium through which the fluid flows, the resin flow can be estimated by Darcy's law:

$$V = -\frac{K}{\mu}\nabla P, \tag{1}$$

where V is the velocity of the flowing resin, ∇P is the applied pressure gradient, μ is the dynamic viscosity of the resin, and K is the permeability tensor. Using a coordinate system aligned to the principal directions of the fiber reinforcement, the permeability tensor can be expressed in the following diagonal form:

$$K = \begin{bmatrix} K_1 & 0 & 0 \\ 0 & K_2 & 0 \\ 0 & 0 & K_3 \end{bmatrix}. \tag{2}$$

Thus, measurements of only three distinct scalar coefficients is required, which can be done according to existing procedures reported in the literature [13–21]. A further simplification is possible: as the thickness of the reinforcement is small compared to its length and width, the problem can be assumed to be 2D. This leaves only the two planar components of permeability, i.e., K_1 and K_2, to be determined. The principal directions of the reinforcement, with respect to which the components of permeability are found, are also found from the experimental tests.

By applying a given pressure gradient, an unsaturated fluid flow is obtained. In the case of a mono-dimensional, i.e., linear, unsaturated flow, Darcy's equation can be expressed as

$$\frac{dx}{dt} = \frac{K_{xx}\Delta P}{\mu\varepsilon x(t)}, \tag{3}$$

where $x(t)$ is the position of the flow front at time t, ΔP is the pressure difference between the inlet and outlet, $\varepsilon = 1 - V_f$ is the void ratio (portion of volume not occupied by the fibres), and K_{xx} is the component of the permeability matrix along the direction of the flow. With the initial conditions $x(t = 0) = 0$, the solution of Equation (3) reads:

$$\frac{x_i^2}{2} = \frac{K_{xx}\Delta P t_i}{\mu\varepsilon}, \tag{4}$$

which allows one to measure the permeability K_{xx}, provided the position of the flow front is tracked throughout the test.

In order to identify the principal directions and the corresponding components of the permeability matrix using linear injection experiments, tests in three planar directions were performed, i.e., 0, 45, and 90° orientations. The principal directions were then found by [22,23]

$$K_1 = K_{\text{exp}}^{0} \frac{\alpha_1 - \alpha_2}{\alpha_1 - \frac{\alpha_2}{\cos(2\beta)}}, \tag{5}$$

$$K_2 = K_{\text{exp}}^{90} \frac{\alpha_1 + \alpha_2}{\alpha_1 + \frac{\alpha_2}{\cos(2\beta)}}, \tag{6}$$

$$\beta = 0.5 \tan^{-1}\left(\frac{\alpha_1}{\alpha_2} - \frac{\alpha_1^2 - \alpha_2^2}{\alpha_2 K_{\text{exp}}^{45}} \right), \tag{7}$$

where the parameters α_1 and α_2 are defined as

$$\alpha_1 = \frac{K_{\text{exp}}^{0} + K_{\text{exp}}^{90}}{2}, \tag{8}$$

$$\alpha_2 = \frac{K_{\text{exp}}^{0} - K_{\text{exp}}^{90}}{2}. \tag{9}$$

2.3. Experimental Setup

The permeability tests were performed on a 1D linear flow test bench. Two different molds were used to test glass and carbon reinforcements (test bench used for glass fibers is shown in Figure 1).

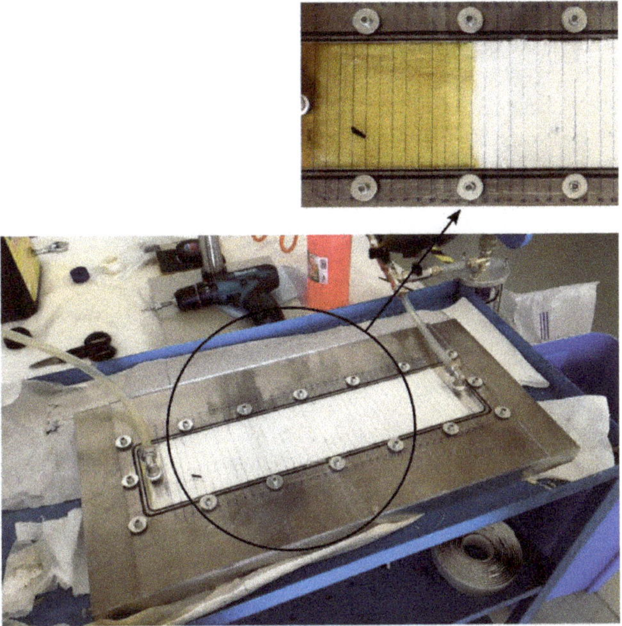

Figure 1. Linear flow test bench used for measurement of glass textile permeability. The oil flow front is clearly visible in the detailed view.

In order to obtain accurate and reproducible results, permeability tests should be carried out with non-reactive fluids: synthetic oil 15W40 and water were used instead of the resin to measure the textile permeability of glass fiber and carbon fiber, respectively. The oil was selected for its controlled viscosity profile over a broad range of temperature, while water is an option in the case of shortage of textiles to be tested (giving the possibility of reuse). The mold cavity had a size of 440 × 100 × 6 mm [16,23–26]. A free length of 20 mm was left at the inlet and outlet on the two sides of the specimen, in order to initiate a stable flow front and prevent potentially dangerous migrations of the fluid to the vacuum apparatus. Thus, a buffer volume was created, which works additionally as a device to reduce pressure fluctuations at the vent. A Plexiglass cover was positioned on top of the mold and fixed using a set of screws; a sealing rubber assured contact between the two mold halves and avoided leakages. Inlet and outlet ports were connected to the top plate for ease of construction.

The mold cavity thickness was chosen in accordance with literature investigations [23,24] and should be considered as an optimum that minimizes two adverse effects. On the one hand, capillarity at the boundaries becomes significant in shallower cavities, resulting in an incorrect reading of the fluid flow; on the other hand, the thickness effects cannot be neglected if thicker cavities are used, thus invalidating the assumption of planar flow.

An adequate stack of textiles should be used to fill the cavity and realize the same fiber volume content as that on the part to be manufactured. Care must be taken when cutting the textiles to avoid wrinkles or voids at the edges of the mold cavity, as these would affect the planar fluid flow. Three repetitions were taken for each direction, i.e., 0°, 45°, and 90° orientations, for a total of 9 trials, for each material. A camera was used to monitor the injection process, and time was measured with a chronometer.

2.4. Numerical Model

A numerical model was developed to simulate the resin infusion process. This consists of modelling the flow of the resin through a porous medium, i.e., the textile. Neglecting the capillarity forces of attraction or repulsion acting at the flow front (as they are deemed sufficiently small in front of the pressure field in RTM), the flow is governed by Darcy's law, e.g., Equation (1). Thus, considering the resin as an incompressible fluid and combining it with Darcy's law result in Richard's equation:

$$\nabla \cdot (\frac{1}{\mu} K \nabla P) = 0. \quad (10)$$

Equation (10) was solved in the software PAM-RTM using the non-conforming finite element method. Two numerical models were developed: The first one simulates the permeability tests of the carbon textile reinforcements, and the second one simulates the infusion process of the entire fiberglass component. The model of the textile permeability test was made up of 52652 3D tetrahedral elements. The pressure difference applied between the inlet and outlet was equal to 40 kPa.

The fiberglass component was modelled with a 2D model in order to diminish the computational cost. A schematic of the model is shown in Figure 2. Four distinct regions can be identified: areas A and B are made of glass fiber textile layers, with one and two textile layers, respectively; region C consists of an empty rectangular channel that is filled by the resin at the beginning of the infusion process; region D is the foam core area. The injection valve is positioned in region C, while the vacuum valve is placed in the middle of D.

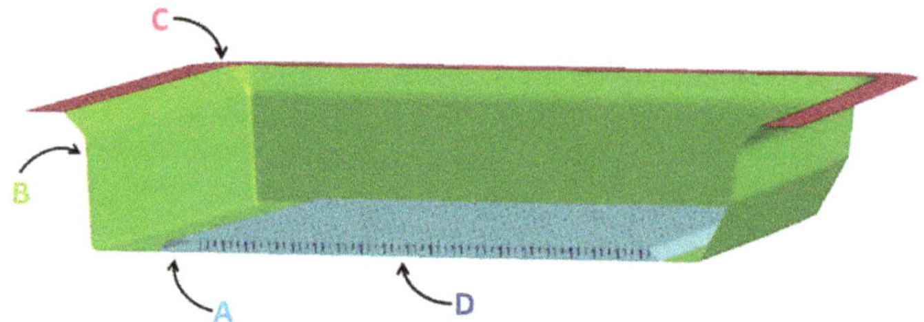

Figure 2. Numerical model description.

The material properties input into the model are given in Table 2. A total of 223,927 triangular elements were used to mesh the body; the core included 862 holes and required a fine mesh with an element size of approximately 2 mm. The empty channel permeability value (1.25×10^{-7} m^2) was used for the permeability of the holes, neglecting the contribution of the glass fibers filling. A sensitivity study was run to substantiate this assumption: by varying the hole permeability between 10^{-3} and 10^{-7} m^2, the filling time changed by less than 2%. This shows that the permeability of the holes has little influence on the filling time of the entire component.

Table 2. Material data of the fiberglass component model.

Zone	K_1 [m^2]	K_2 [m^2]	β [deg]	Thickness [mm]	Fiber Content
A	2.530×10^{-9}	2.074×10^{-9}	35	3.2	0.328
B	2.530×10^{-9}	2.074×10^{-9}	35	6.4	0.328
C (edges and injection area)	14.2×10^{-6}	14.2×10^{-6}	-	7 (edges) 1 (injection area)	0
D (core Holes)	1.25×10^{-7}	1.25×10^{-7}	-	1	0

A pressure of 101.3 kPa was applied at the injection valve, while the pressure at the vent, i.e., in the middle of region D, was set equal to 5 kPa.

3. Results and Discussion

3.1. Material Characterization Tests

The values of the resin and test oil viscosity measured in the experimental tests are given in Table 3. Clearly, as the room and production temperatures are almost equal, the resin viscosity is almost unaffected by these temperature changes.

Table 3. Resin and test oil viscosity as measured in experimental tests.

	Orthophthalic Resin		**Test Oil**	
Rotational Speed [rpm]	**μ at 26.6 °C [cPs]**	**μ at 31 °C [cPs]**	**μ at 26.6 °C [cPs]**	**μ at 31 °C [cPs]**
20	196	199	181	152
30	200	199	182	151
50	209	203	187	154
100	233	226	203	175
Mean Value μ	**210**	**207**	**188**	**158**

Table 4 shows the results of the permeability tests, grouped according to their orientation. As visible from the table, the scatter in the data is limited: for every orientation, the results are within a 5% error range.

Table 4. Glass textile permeability measured in the tests.

Permeability	K^0_{exp} **[10^{-9} m^2]**	K^{45}_{exp} **[10^{-9} m^2]**	K^{90}_{exp} **[10^{-9} m^2]**
Trial 1	2.400	2.119	2.126
Trial 2	2.336	2.127	2.285
Trial 3	2.335	2.006	2.207
Mean Value	**2.357**	**2.084**	**2.206**
St. Deviation	0.030	0.055	0.065

By using Equations (5)–(7), the principal textile permeabilities can be computed. This gives the values collected in Table 5.

Table 5. Principal permeabilities of the glass textile.

	K_1 **[10^{-9} m^2]**	K_2 **[10^{-9} m^2]**	**β [deg]**
Principal permeability	2.530	2.074	−35

Knowing the values of the principal permeabilities and their corresponding orientation, it is possible to compute the effective permeability K_{eff} along a general direction θ, which is defined as

$$\sqrt{K_{eff}} = \frac{\sqrt{K_1 K_2}}{K_1 \sin(\theta) K_2 \cos(\theta)}. \tag{11}$$

This allows one to visualize the flow distribution in the given layup as a function of the orientation angle θ. For the tested glass textile layup, the resulting effective permeability is plotted in Figure 3. Effective permeability of the fiberglass textile plotted as a function of the orientation angle θ.

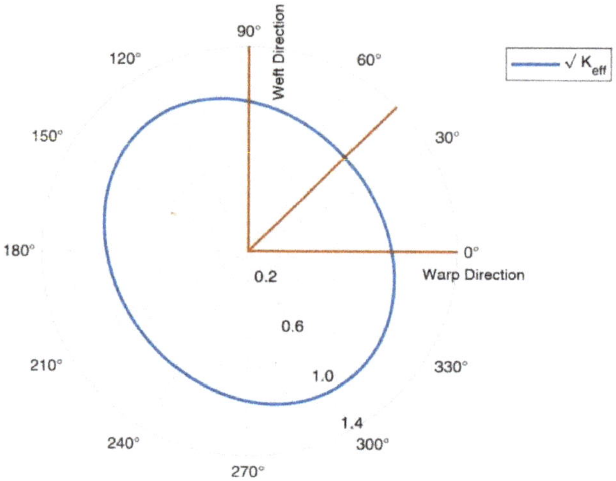

Figure 3. Effective permeability of the fiberglass textile plotted as a function of the orientation angle θ.

As for the carbon fiber reinforcements, the permeability of the three different carbon fiber reinforcements was tested, i.e., (ply A) unidirectional fiber 12K T700 300 g/m^2, (ply B) triaxial fiber (30/90/−30) 24K T700 300 g/m^2, and (ply C) Twill 2 × 2 fabric 12K T700 630 g/m^2.

Three tests were performed for each textile. Data obtained are listed in Tables 6 and 7:

Table 6. Local permeability of the carbon textile.

Plies Type		K^0_{exp} [10^{-9} m^2]	K^{90}_{exp} [10^{-9} m^2]	K^{45}_{exp} [10^{-9} m^2]
A	**Mean value**	0.10	0.06	0.04
	St. Deviation	0.008	0.003	0.004
B	**Mean value**	0.05	0.07	0.10
	St. Deviation	0.008	0.006	0.016
C	**Mean value**	0.18	0.16	0.14
	St. Deviation	0.027	0.010	0.002

Table 7. Principal permeability of the carbon textile.

Plies Type	K_1 [10^{-9} m^2]	K_2 [10^{-9} m^2]	β [deg]
A	0.10	0.04	5.6
B	0.10	0.05	85.0
C	0.18	0.14	0.7

Data show that the permeabilities of the selected carbon fiber textile styles are at least one order of magnitude lower than those of glass fiber. The achievement was expected as the glass fiber textile comprises a multi-layer of 600 g/m^2 E-glass biaxial fibers [+/−45° orientation] and a 450 g/m^2 chopped strand mat stitched by means of a 18 g/m^2 PP for enhancing permeability characteristics. In addition, it is shown that anisotropic permeability occurs even in balanced textiles where no apparent preferential direction is expected. Under deeper analysis, the ratio between major and minor in-plane permeability spans between 1.2 for the glass textile and 2.0 for carbon triaxial fiber (ply B) and 1.3 for the carbon 2 × 2 twill. These results indicate that proper characterization of the textile permeability is crucial in the design phase of an infusion process, as the filling path and time can be significantly affected by the fabric configuration and layup.

3.2. Numerical Models and Experimental Comparison

The average filling time measured in the carbon textiles permeability tests is reported in Table 8, together with values computed by the numerical model. Comparison between the experimental and numerical results shows a limited difference within experimental data scatter. For better visualization of the numerical data, they are also plotted in Figure 4. as a contour map. Clearly, the points located further from the injection valve, i.e., the right-most point in the figure, have the highest filling time.

Table 8. Carbon textile reinforcement permeability experimental and numerical filling time results.

		Ply A 0°	Ply A 90°	Ply B 0°	Ply B 90°	Ply C 0°	Ply C 90°
Exp. [s]	Mean value	2.5	6.9	2.7	5.0	3.0	11.0
	St. Dev.	0.2	0.7	0.5	0.9	0.3	1.6
Num. [s]		2.7	7.5	2.9	5.4	3.1	9.5

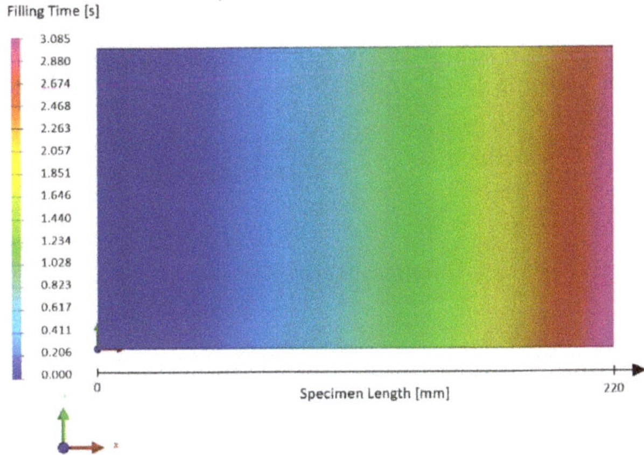

Figure 4. Contour map of the filling time computed numerically: example of case Ply C 0° (color scale in s).

For the fiberglass component, the filling time calculated in the simulation is equal to 304.3 s and the resin flow is uniform in the entire component. Figure 5 shows the position of the flow front at four time instants, while a contour plot of the filling time is given in Figure 6.

As visible in Figure 5, the resin flows through all the empty borders in less than 1 s. After that, the flow front uniformly follows the vertical faces and the flow eventually forms a circular shape around the vacuum valve.

Therefore, the analysis of the flow pattern in the numerical simulation shows that the injection and vent valves are in good positions to assure an adequate filling path of this simple component.

The filling time of the real component was measured to be 303 s. A comparison of this value to that predicted by the numerical model shows an error of 0.44%.

Moreover, by analyzing the component pulled out of the mold after resin polymerization (Figure 7), there was an absence of holes, bubbles, or other defects. This shows that, also in the real condition, valve locations and pressure settings are adequate in order to obtain a good structure.

Therefore, comparing the numerical and experimental results shows a good correspondence in all analyzed cases. This highlights the reliability of the methodology discussed in this paper. In fact, it has been proved that the RTM process simulation, by means of the PAM-RTM software, could give reliable results for optimization of the injection strategy in complex components, provided that realistic values are input to the model. To this end, accurate material characterization is needed, which must take into account the dependence of the material properties on the specific experimental conditions (e.g., temperature). Moreover, the results presented are in good accordance with other studies where good correspondence between RTM (as well as VARI and C-RTM) modelling and experimental results was obtained [1–3,27,28]. In summary, a workflow explaining the phases and procedures for the characterization of the materials and then to simulate the RTM process, as followed in this paper, is reported in Figure 8.

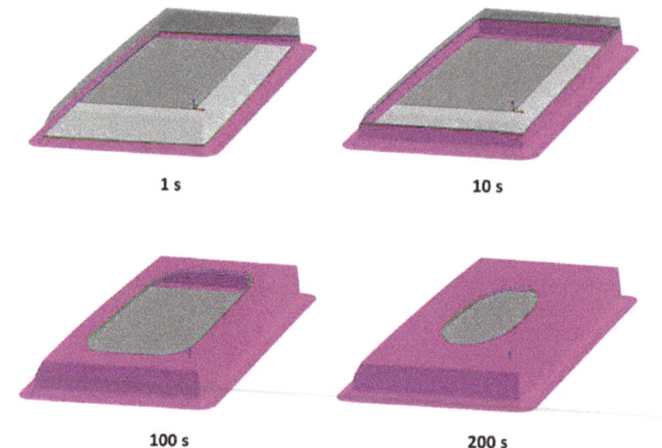

Figure 5. Visualization of the flow front calculated by the numerical model.

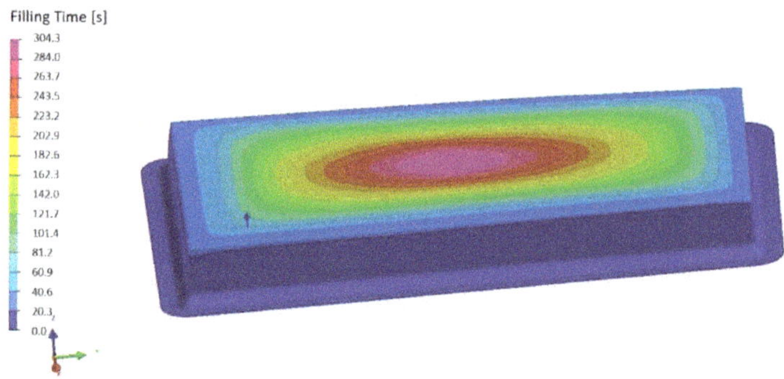

Figure 6. Contour plot of the filling time computed by the numerical model.

Figure 7. Fiberglass component manufacturing: (**a**) Dry material before resin infusion, (**b**) finished part.

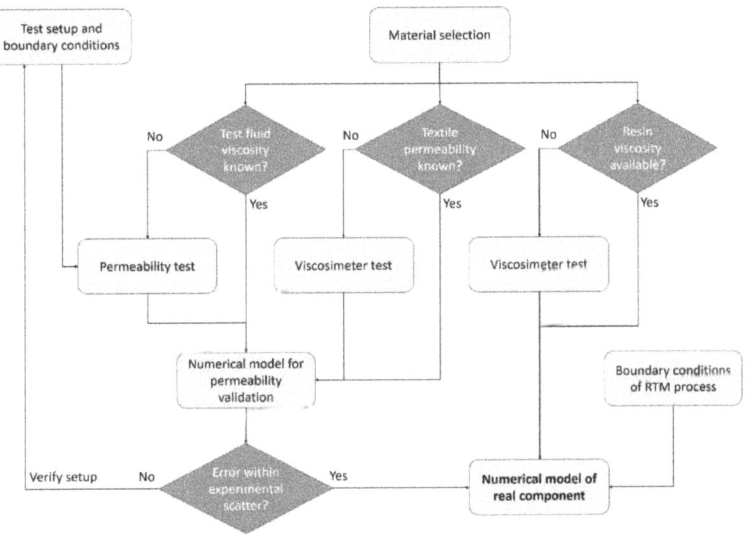

Figure 8. Workflow to model the resin transfer molding (RTM) process of a real component.

4. Conclusions

In this paper, the procedure for implementing reliable inputs in the simulation of an RTM-based process is presented and discussed. The experimental setup for textiles permeability characterization is described, as well as the procedure for retrieving local and global permeability. Four different types of textiles (three carbon fiber styles and one glass fiber multilayer textile) have been characterized in order to provide to readers reference data for different types and styles of textiles. The permeabilities of the carbon fiber textiles were found to be one order of magnitude lower than that of glass fiber, whereas fiber alignment had a less relevant influence. Moreover, tests show that anisotropic permeability occurs even in balanced textiles where no apparent preferential direction is expected. The characterization of the viscosity of the orthophthalic resin Lavesan LERPOL 666/S and of the synthetic oil 15W40 showed comparable values at room temperature.

ESI PAM-RTM software was used to perform RTM simulations. As a first step, simulation of permeability characterization in the different directions was performed by means of 3D models. Filling time results showed an average good correspondence with experimental data (3.6% average error). As a second step, the resin infiltration of a 15.85 m (52 ft) sailing yacht component, comprised of stitched glass fibers around a perforated core, was implemented in a 2D model. Output data including filling flow and time proved consistent (0.44% error in filling time) with data acquired in industrial environment, validating the procedure proposed in this paper.

Author Contributions: Conceptualization, L.D. and F.R.; methodology, L.D.; software, M.P.F., L.G. and M.G.; validation, M.P.F., E.T. and N.Z.; formal analysis, M.P.F. and F.R.; investigation, F.R.; data curation, L.D.; writing—original draft preparation, M.P.F.; writing—review and editing, N.Z and E.T. All authors have read and agreed to the published version of the manuscript.

Funding: This research received no external funding.

Acknowledgments: The authors thank Corset&Co Srl for the naval component production.

Conflicts of Interest: The authors declare no conflict of interest

References

1. Poodts, E.; Minak, G.; Mazzocchetti, L.; Giorgini, L. Fabrication, process simulation and testing of a thick CFRP component using the RTM process. *Compos. Part B Eng.* **2014**, *56*, 673–680. [CrossRef]
2. Rondina, F.; Taddia, S.; Mazzocchetti, L.; Donati, L.; Minak, G.; Rosenberg, P.; Bedeschi, A.; Dolcini, E. Development of full carbon wheels for sport cars with high-volume technology. *Compos. Struct.* **2018**, *192*, 368–378. [CrossRef]
3. Poodts, E.; Minak, G.; Dolcini, E.; Donati, L. FE analysis and production experience of a sandwich structure component manufactured by means of vacuum assisted resin infusion process. *Compos. Part B Eng.* **2013**, *53*, 179–186. [CrossRef]
4. Young, W.B.; Rupel, K.; Han, K.; Lee, L.J.; Liou, M.J. Analysis of resin injection molding in molds with preplaced fiber mats. *II: Numerical simulation and experiments of mold filling. Polym. Compos.* **1991**, *12*, 30–38.
5. Liu, B.; Bickerton, S.; Advani, S.G. Modelling and simulation of resin transfer moulding (RTM)—Gate control, venting and dry spot prediction. *Compos. Part A Appl. Sci. Manuf.* **1996**, *27*, 135–141. [CrossRef]
6. Trochu, F.; Ruiz, E.; Achim, V.; Soukane, S. Advanced numerical simulation of liquid composite molding for process analysis and optimization. In *Proceedings of the Composites Part A: Applied Science and Manufacturing*; Elsevier Ltd.: Amsterdam, The Netherlands, 2006; Volume 37, pp. 890–902.
7. Grössing, H.; Stadlmajer, N.; Fauster, E.; Fleischmann, M.; Schledjewski, R. Flow front advancement during composite processing: Predictions from numerical filling simulation tools in comparison with real-world experiments. *Polym. Compos.* **2016**, *37*, 2782–2793. [CrossRef]
8. Dereims, A.; Drapier, S.; Bergheau, J.M.; De Luca, P. 3D robust iterative coupling of Stokes, Darcy and solid mechanics for low permeability media undergoing finite strains. *Finite Elem. Anal. Des.* **2015**, *94*, 1–15. [CrossRef]
9. Celle, P.; Drapier, S.; Bergheau, J.M. Numerical modelling of liquid infusion into fibrous media undergoing compaction. *Eur. J. Mech. A Solids* **2008**, *27*, 647–661. [CrossRef]
10. Imbert, M. High Speed Reactive Resin Transfer Moulding (RTM) Process Simulation for Mass Production of Automotive Structural Parts. *SAE Int. J. Mater. Manuf.* **2015**, *8*, 503–515. [CrossRef]
11. Dereims, A.; Chatel, S.; Marquette, P.; Dufort, L. Accurate Liquid Resin Infusion simulation through a Fluid-Solid coupled approach | ESI Group. In Proceedings of the SAMPE., Seattle, WA, USA, 22–25 May 2017.
12. Naik, N.K.; Sirisha, M.; Inani, A. Permeability characterization of polymer matrix composites by RTM/VARTM. *Prog. Aerosp. Sci.* **2014**, *65*, 22–40. [CrossRef]
13. Lundström, T.S.; Stenberg, R.; Bergström, R.; Partanen, H.; Birkeland, P.A. In-plane permeability measurements: A nordic round-robin study. *Compos. Part A Appl. Sci. Manuf.* **2000**, *31*, 29–43. [CrossRef]

14. Weitzenböck, J.R.; Shenoi, R.A.; Wilson, P.A. Radial flow permeability measurement. *Part A: Theory. Compos. Part A Appl. Sci. Manuf.* **1999**, *30*, 781–796. [CrossRef]
15. Weitzenböck, J.R.; Shenoi, R.A.; Wilson, P.A. Measurement of three-dimensional permeability. *Compos. Part A Appl. Sci. Manuf.* **1998**, *29*, 159–169. [CrossRef]
16. Bodaghi, M.; Lomov, S.V.; Simacek, P.; Correia, N.C.; Advani, S.G. On the variability of permeability induced by reinforcement distortions and dual scale flow in liquid composite moulding: A review. *Compos. Part A Appl. Sci. Manuf.* **2019**, *120*, 188–210. [CrossRef]
17. Sharma, S.; Siginer, D.A. Permeability measurement methods in porous media of fiber reinforced composites. *Appl. Mech. Rev.* **2010**, *63*, 1–19. [CrossRef]
18. Ince, M.E. Air permeability characterization of glass fiber nonwoven fabric for liquid composite molding applications. In *Proceedings of the IOP Conference Series: Materials Science and Engineering*; Institute of Physics Publishing: Bristol, UK, 2018; Volume 459.
19. Ya'Acob, A.M.; Razali, D.A.; Anwar, U.A.; Radhi, A.H.; Ishak, A.A.; Minhat, M.; Mohd Aris, K.D.; Johari, M.K.; Casey, T. Preliminary Study on GF/Carbon/Epoxy Composite Permeability in Designing Close Compartment Processing. In *Proceedings of the IOP Conference Series: Materials Science and Engineering*; Institute of Physics Publishing: Bristol, UK, 2018; Volume 370.
20. Kim, J.I.; Hwang, Y.T.; Choi, K.H.; Kim, H.J.; Kim, H.S. Prediction of the vacuum assisted resin transfer molding (VARTM) process considering the directional permeability of sheared woven fabric. *Compos. Struct.* **2019**, *211*, 236–243. [CrossRef]
21. Becker, D.; Mitschang, P. Measurement System for On-Line Compaction Monitoring of Textile Reaction to Out-of-Plane Impregnation. *Adv. Compos. Lett.* **2014**, *23*. [CrossRef]
22. Demaría, C.; Ruiz, E.; Trochu, F. In-plane anisotropic permeability characterization of deformed woven fabrics by unidirectional injection. Part I: Experimental results. *Polym. Compos.* **2007**, *28*, 797–811.
23. Arbter, R.; Beraud, J.M.; Binetruy, C.; Bizet, L.; Bréard, J.; Comas-Cardona, S.; Demaria, C.; Endruweit, A.; Ermanni, P.; Gommer, F.; et al. Experimental determination of the permeability of textiles: A benchmark exercise. *Compos. Part A Appl. Sci. Manuf.* **2011**, *42*, 1157–1168. [CrossRef]
24. Vernet, N.; Ruiz, E.; Advani, S.; Alms, J.B.; Aubert, M.; Barburski, M.; Barari, B.; Beraud, J.M.; Berg, D.C.; Correia, N.; et al. Experimental determination of the permeability of engineering textiles: Benchmark II. *Compos. Part A Appl. Sci. Manuf.* **2014**, *61*, 172–184. [CrossRef]
25. Ferland, P.; Guittard, D.; Trochu, F. Concurrent methods for permeability measurement in resin transfer molding. *Polym. Compos.* **1996**, *17*, 149–158. [CrossRef]
26. Lundström, T.S.; Gebart, B.R.; Sandlund, E. In-plane permeability measurements on fiber reinforcements by the multi-cavity parallel flow technique. *Polym. Compos.* **1999**, *20*, 146–154. [CrossRef]
27. Dereims, A.; Zhao, S.; Yu, H.; Pasupuleti, P.; Doroudian, M.; Rodgers, W.; Aitharaju, V. Compression Resin Transfer Molding (C-RTM) Simulation Using a Coupled Fluid-solid Approach | ESI Group. In Proceedings of the American Society for Composites 32nd Technical Conference, West Lafayette, IN, USA, 23–25 October 2017.
28. Marquette, P.; Dereims, A.; Ogawa, T.; Kobayashi, M. Numerical Methods For 3D Compressive RTM Simulations | ESI Group. In Proceedings of the ECCM17, Munich, Germany, 26–30 June 2016.

Article

Investigation of Electrodeposition External Conditions on Morphology and Texture of Ni/SiC*w* Composite Coatings

Liyan Lai, Hongfang Li, Yunna Sun, Guifu Ding *, Hong Wang and Zhuoqing Yang

National Key Laboratory of Science and Technology on Micro/Nano Fabrication, School of Electronic Information and Electrical Engineering, Shanghai Jiao Tong University, Shanghai 200240, China
* Correspondence: gfding@sjtu.edu.cn; Tel.: +86-21-34206686

Received: 19 August 2019; Accepted: 9 September 2019; Published: 12 September 2019

Abstract: Recently, a strategy of synthesizing SiC whisker-reinforced nickel (Ni/SiC*w*) composites with excellent mechanical properties by electrodeposition has been proposed for exploring its potential applications in micromechanical devices. In this paper, a series of external conditions that affected the content of SiC whiskers in composite films were studied, such as cathode current density, stirring rate and electrolyte temperature. The experimental results indicated that the optimum morphology was obtained at a stirring speed of 300 rpm, a temperature of 50 °C, and a current density of 18 mA/cm^2. Additionally, the content of SiC whiskers and textural preference were also investigated by varying its external conditions, and the results demonstrated that the composites with high mass percentage whiskers are more advantageous for electrocrystallization of Ni in the (200) orientation. Finally, the relationship between external conditions and intrinsic morphology, composition and texture of Ni/SiC*w* composites was revealed, and it provides a constructive approach to fabricate the high-content SiC whiskers of these composites.

Keywords: physical conditions; electrodeposition; SiC whisker; texture; morphology

---

1. Introduction

Electrodeposition is one of the most commercially successful and inexpensive superior techniques for fabrication of metallic coatings. However, the conventional electrodeposition methods are very simple, since it has only one phase throughout the process. The pure electroplated metals are difficult to satisfy the current requirements, due to their plastic deformation. In order to resolve this problem, the method of co-deposition insoluble solid particles in metal matrix was established and recognized for fabrication the composite coatings. The reinforcing particles, suspending in the electrolyte, are entrapped and incorporated into the metal matrix by adopting suitable methods during the co-deposition process, such as electrophoresis, adsorption or mechanical entrapment [1]. In 1928, Fink et al. [2] introduced the co-deposition method to produce Cu/graphite composite coatings that applied in car engines for the first time. Besides, the first patent of co-deposition was issued by Grazen, in 1962 [3]. Since then, considerable researches have been focused on different types of composites fabricated by the co-deposition method. These composites include ceramics, metals, polymer or microcapsule/liquid incorporated into metal matrices, which could improve the coating properties as mechanical strength, wear/friction-resistance, high-temperature oxidation-resistant and corrosion-resistance [4–15]. Based on the SiCw with features in high modulus and tensile strength, good wear resistance, as well as dimensional stability, recently, we have successfully synthesized high mechanical strength Ni/SiCw composites by electrodeposition.

Since the performance of composite coatings is highly sensitive to external conditions during the co-deposition process, such as current density, temperature, stirring speed, and current density, etc.

Therefore, numerous efforts have been dedicated to improving its microstructure and micromorphology of the deposited coatings for enhancing their properties. Lately, relevant studies have revealed that the content of particles in the composite coating can be controlled by the external conditions during the deposition process, and resulting in a difference in the performance of the composite coatings. H. Gül et al. [16] reported the nickel metal matrix composites reinforced with SiC submicron particles by electrodeposition. Besides, the influence of stirring speed and particles concentration on co-deposited composite coatings was investigated. The results indicated that the maximum of particle concentration was 20 g/L and the high stirring speed was chosen to obtain the wear resistant nickel matrix coatings. P. Gyftou et al. [17] applied both direct and pulse current conditions to produce Ni/ nano-SiC composites, and it proved that pulse electrodeposition significantly improved the hardness of the Ni/SiC composite deposits. C.K. Chung et al. [18] investigated the electrodeposition process of Ni at low electrolytic temperatures (5–20 °C). It demonstrated that low temperature could enhance the hardness up to 6.18 GPa and produce the smoothness of films. Tushar Borkar et al. [19] presented the effects of the deposition conditions and the nanoparticle concentration in the electrolyte on the surface microstructure, crystallographic micro-texture, microhardness, and tribological properties of coatings. The reinforcement of Ni-Al_2O_3 nanoparticles significantly improved microhardness and wear resistance of the composite coatings. Based on the previous studies, it is necessary to clearly understand the influence of various conditions on the co-deposition of SiC whiskers in the nickel matrix, so as to consciously control the whiskers content and effectively manipulate the performance of the composite coatings. The purpose of this study is to achieve high SiC whiskers content in Ni/SiC*w* composite coatings. In addition, the influence of the parameters, such as stirring speed, electrolyte temperature and cathode current density on the SiC*w* content in composites were investigated, which clearly indicated the relationship between the external conditions and the content of SiC whiskers in the composites.

2. Materials and Preparation

The SiC whiskers (β- SiC*w*, XFNANO Material Co., Ltd., Nanjing, China), with an average length of 50 μm and diameter of 0.2 μm, were selected. Ni/SiC*w* composite coatings were prepared by a constant-current electrodeposition method from a nickel sulfamate electrolyte containing SiC whiskers as a reinforcement phase. Prior to the SiC*w* addition into the electroplating solution, it was first stirred in 3 vol. % hydrofluoric acid and refluxed for 8 h at 95 °C [20], then ultrasonically washed in distilled water until the pH becomes neutral, subsequently modified by γ-aminopropyltriethoxysilane (KH550). These treatments can remove the SiC*w* surface impurities (SiO_2) and improve its wettability, as well as dispersibility. After the above treatments, SiC whiskers were added in the electrolyte by continuous magnetically stirring with a rate of 300 rpm for at least 24 h, which can prevent whiskers agglomeration and maintain them in suspension state. According to a number of optimization experiments, the optimal whisker concentration in the electrolyte was 0.8 g/L. The electrolyte composition and electrodeposition parameters are listed in Table 1. Analytical reagents and deionized water were used to prepare the plating solution. After electrodeposition, Ni/SiC*w* composite coatings were ultrasonically cleaned in distilled water for 10 min to remove loosely adsorbed SiC*w* from the surface.

The surface morphology and microstructure of the composite coatings were characterized by scanning electron microscopy (SEM; ULTRA55, Zeiss, Germany). The amount of embedded SiC*w* was evaluated by the energy dispersive X-ray spectroscopy (EDS) with the same system of SEM. The phase structure analysis of the coatings was conducted by X-Ray Diffraction (XRD, D8 Advance, Bruker-axs) operating with Cu Kα (λ = 1.54178 Å) radiation at room temperature. The preferred XRD orientation index TC(*hkl*) was calculated, and the texture co-efficient (TC) for each (hkl) reflection is given by [21,22]:

$$TC(hkl) = \frac{I_{(hkl)}/I_{0(hkl)}}{\frac{1}{n}\sum I_{(hkl)}/I_{0(hkl)}} \quad (1)$$

where TC (*hkl*) is the texture coefficient of the specific (*hkl*) plane; I (*hkl*) and I_0 (*hkl*) are the diffraction intensity of the electrodeposited nickel and the powder pattern in the JCPDS cards, respectively; n is the number of reflection peaks used in the calculation. Three peaks of (111), (200) and (221) are used to calculate the texture coefficient.

Table 1. Operating parameters for Ni/SiC*w* composite electroplating.

Electrolyte Composition		Electrodeposition Parameters	
$Ni(NH_2SO_3)_2{\cdot}4H_2O$	300 g/L	Current density	2~40 mA/cm^2
$NiCl_2{\cdot}6H_2O$	40 g/L	Temperature	20~60 °C
H_3BO_3	30 g/L	Magnetic stirring	0~500 rpm
SiC whisker	0.8 g/L	pH	4.1

3. Results and Discussions

3.1. Effects of Electrodeposition Parameters on Morphology and Composition of the Ni/ SiCw Composite

Figure 1 shows the surface morphology of various current density for the Ni/SiC*w* composite coatings, which was prepared at a stirring speed of 300 rpm, a temperature of 50 °C, and a current density from 2 mA/cm^2 to 40 mA/cm^2. As shown in Figure 1, obviously, the microstructure and content of SiC*w* in coatings were appreciably affected by the current density. When the current density was lower than 18 mA/cm^2, the glossy surface of the nickel matrix with uniformly distributed SiC whiskers could be seen, and the content of SiC*w* increased with the increase of current density (Figure 1a–d). However, when the current density was higher than 25 mA/cm^2, the whiskers content reduced, and the surface morphology of coatings became nodular and inhomogeneous (Figure 1e–f). These phenomena made clear that the electrodeposits prepared at low current density tend to form smooth and luminous surface morphology with fine grains.

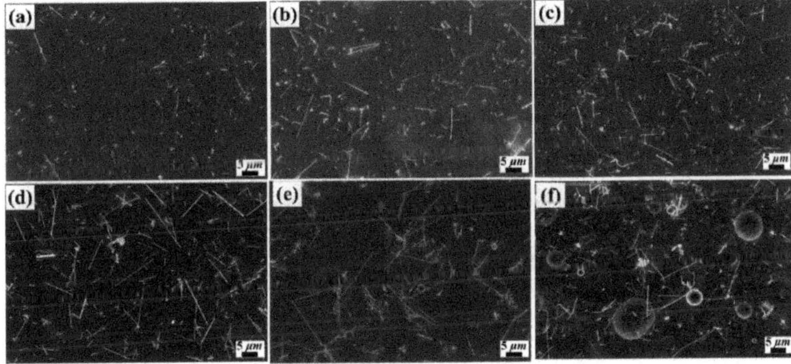

Figure 1. SEM images of Ni/SiC*w* composite coating fabricated from the electrolyte at stirring speed of 200 rpm, and temperatures of 35 °C, and current density of (**a**) 2 mA/cm^2, (**b**) 6 mA/cm^2, (**c**) 10 mA/cm^2, (**d**) 18 mA/cm^2, (**e**) 25 mA/cm^2, (**f**) 40 mA/cm^2.

The influences of current density on composite coatings could be explained by the following aspects. Firstly, as the cathode current density increases, the Coulomb force between the whisker-adsorbed metal ions and the cathode increases, which can increase the deposition rate of the matrix metal and save time for SiCw to be embedded in Ni matrix. That is, the content of SiCw in the nickel matrix increases at per unit time. Secondly, as the cathode current density increases, the cathode overpotential and the electric field force increases correspondingly, which facilitates the electrostatic attraction of the cathode to the solid particles adsorbing the positive ions. At higher overvoltage, the transportation of metal ions to cathode becomes an important factor in electrodeposition, and it is well known that the addition of inert particles could enhance this transportation [23]. Compared with lower current density, the rate of whisker that transfer to cathode surface and embed into the composite coatings increases remarkably when the cathode current density is too high. Thirdly, part of the surface of the cathode is covered by whiskers, due to its embedding behavior, which results in the increase of the cathode overpotential, the enhancement of hydrogen adsorption capacity and the formation of alkaline salts near the cathode [4]. These by-products that adsorbed on the cathode surface or embedded into the coating also prevent the whisker from co-depositing with nickel metal and lead to a decrease of the SiCw content in electrodeposits. Besides, at a higher overpotential, the dendrites or nodules were formed on the coating's surface, due to the discharge of metal ions near the cathode, which cause the discharge at the swelling of deposits and it is consistent with previous literature reported by other researchers [2].

During the electrodeposition process, the deposited crystal and grew behavior are closely related to the electrolyte temperature, which plays an essential role in controlling its texture and grain size. Moreover, the mass transfer process of metal ions and whiskers to the cathode surface could be enhanced at an appropriate electrodeposition temperature [24]. Therefore, it is indispensable to evaluate the effect of electrolyte temperature on the composite for the purpose of obtaining the high content of SiCw in the coating with smooth and uniform morphology. In this experiment, the temperature of electrolyte ranging from 20 °C to 60 °C were investigated at a stirring speed of 200 rpm and a current density 18 mA/cm^2. As illustrated in Figure 2, the dendrites and nodules could be observed clearly at a lower temperature (Figure 2a), while the smooth and luminous surface accompanied by the increasing whisker content was obtained at a temperature rising to 50 °C (Figure 2d).In addition, the nickel grains became finer, and the composite coatings were also strengthened after adding the whiskers. However, the whiskers tended to aggregate and decreased its content in composite coatings when the temperature was exceeding 50 °C, which resulted in the deposited grains was coarse and oversized.

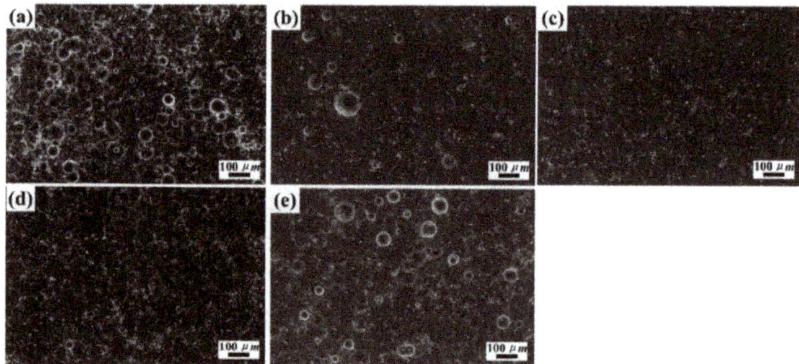

Figure 2. SEM images of Ni/SiC*w* composite coating fabricated from the electrolyte at stirring speed of 200 rpm, and current density of 18 mA/cm^2, and temperatures of (**a**) 20 °C, (**b**) 30 °C, (**c**) 40 °C, (**d**) 50 °C, (**e**) 60 °C.

In general, both the average kinetic energy of ions and the viscosity of electrolytic could be consolidated by increasing the electrolyte temperature, which all consequently facilities the transportation of whiskers. Besides, the diffusion mobility of metal ions adsorbed on the cathode surface would be increased on the electrode surface, which resulted in an increase of the whiskers contents in composites coating. Whereas, the anions diffusion rate, the overpotential and the electric field force of the cathode reduced when the temperature reached to a specific value, which was detrimental to the whiskers embedded into the nickel matrix. In addition, the electrolyte would evaporate seriously at a higher temperature, thus, affecting the deposition of active ingredients onto the cathode and coursing the surface of composites. As a result, the excessively high temperature led to a reduction of the SiC*w* adsorption ability and the cathode electric field force, which ultimately reduced the SiCw content in the composite coatings.

In order to prevent the precipitation and agglomeration of SiCw in the electrolyte, mechanical stirring was adopted as a key technology in electroplating process, due to the high surface energy of the SiCw. Particularly, the content of particles in composite coatings is depended largely on the strength of stirring, which is advantageous to the uniform distribution of particles in electrolyte and the successful transportation to cathode surface [23]. Figure 3 shows the SEM images relating to the morphology of Ni/SiCw coatings fabricated at a different stirring speed. The conditions of current density and temperature were 18 mA/cm^2 and 50 °C, and the stirring speed was controlled at 0, 100, 300, and 500 rpm, respectively. Due to the gravity, a large number of whiskers accumulated and covered the major cathode surface without stirring, which resulted in an increase of the cathode overpotential and a decrease of the discharged metal ions near the cathode. Therefore, metal ions discharged at the corner and appeared dendrites or nodules in the composite coatings under the condition without stirring. With the increasing strength of stirring, the mass transfer rate of electrolyte and the effective concentration of whisker were also increased accordingly, which could promote the transportation of SiC*w* and increase the whisker content in the composite coatings. However, the electrolyte flew at a high speed when the stirring strength is higher than 300 rpm, which caused intense shock to the cathode surface, led to high-speed transfer of the electrolyte accompanied by whiskers and suppressed the attachment of whiskers to the cathode surface. More seriously, whiskers that were not fully embedded in the nickel matrix might be released from the cathode surface to the electrolyte, which resulted in a dramatic decrease of the whisker content in the composite coatings.

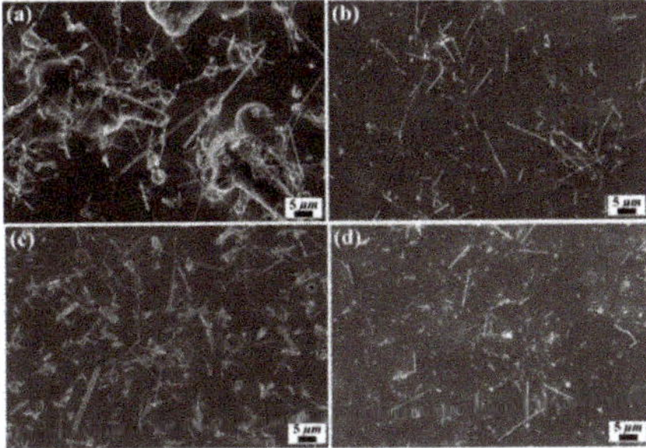

Figure 3. SEM images of Ni/SiC*w* composite coating fabricated from the electrolyte at a current density of 18 mA/cm^2, and temperatures of 50 °C, and stirring speed of (**a**) 0 rpm, (**b**) 100 rpm, (c) 300 rpm, (**d**) 500 rpm.

The cross-sectional SEM images of Ni/SiCw composite coatings prepared with different electrodeposited conditions were shown in Figure 4. From Figure 4a,b, in which the current density ranges from 2 to 8 mA cm^{-2}, an increase in the number of SiCw (pointed by an arrow), which are fairly well dispersed in the nickel matrix, can be clearly observed on the fracture surface. According to the micrographs of the cross section, the fracture surface does not represent the initial interface between the the nickel matrix and SiCw after breaking with external force, so the interface was further illustrated after treatment by an ion beam thinner (Figure 4c,d). From the cross-sectional images of the thinned surface, the whiskers were cut in an identical plane, but in different orientations, due to their random distribution in the the nickel matrix; meanwhile, multiple shapes of whiskers after thinning by ion beam thinner were also noticed.

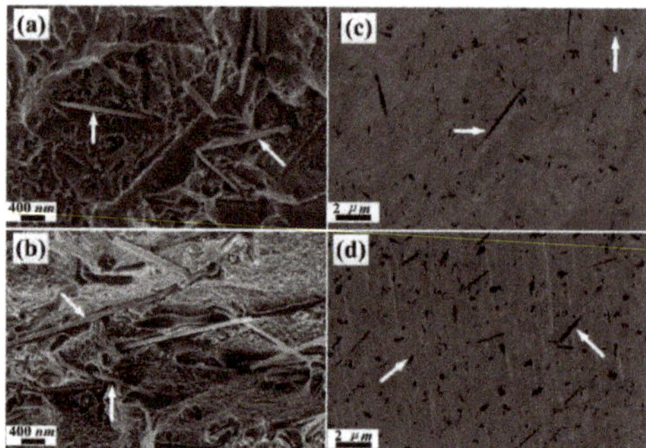

Figure 4. SEM images of the cross section for the fracture (**a**,**b**) and thinned surface (**c**,**d**) of the electrodeposited Ni/SiCw composite coatings (**a**) SiCw 0.8 g L^{-1}, 2 mA/cm^2, 35 °C, 150 rpm; (**b**) SiCw 0.8 g L^{-1}, 18 mA/cm^2, 35 °C, 150 rpm; (**c**) and (**d**) represent the thinned surface of (**a**) and (**b**), respectively.

The EDS spectra of Ni/SiCw composite coatings, prepared at different conditions, are illustrated in Figure 5; and is related to the EDS spectra, which has been listed, as shown in Table 2. As shown in Figure 5a, only Ni peaks were observed in the sample without SiCw coating, whereas the peaks of Ni, Si and C were all collected in Figure 5b–e. The EDS results indicated that the SiC whiskers were incorporated into the nickel matrix successfully. Besides, the obviously increased content of Si and C could be seen from Figure 5b,c, which was resulted from the increased current density from 2 to 18 mA/cm^2. Since the co-deposition rate of SiCw was strongly related to that of Ni matrix in the process of electrodeposition, then increasing current density could promote the deposition rate of matrix metals and save the time of SiCw embedding into the Ni matrix. In Figure 5c,d, the EDS spectra of the Ni/SiCw coatings deposited at 35 °C and 50 °C, which showed that the higher temperature facilitated the increase of whiskers content in electrodeposits. It shows that the higher temperature facilitates the increase of whiskers content in electrodeposits. As to the influence of stirring speed on the content of SiC*w*, the stronger stirring speed can contribute to the whiskers transmitting to the cathode surface from the electrolyte which results in its higher content in coatings (Figure 5e).

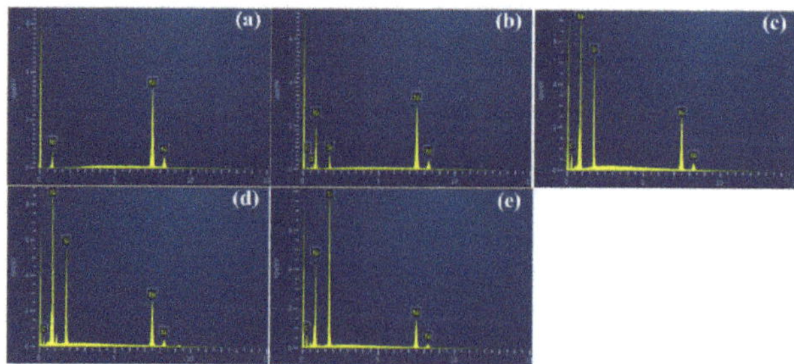

Figure 5. EDS spectra of electrodeposited Ni/SiC*w* composite coatings (**a**) without SiC*w*, 2 mA/cm^2, 35 °C, 150 rpm; (**b**) SiC*w* 0.8 g L^{-1}, 2 mA/cm^2, 35 °C, 150 rpm; (**c**) SiC*w* 0.8 g L^{-1}, 18 mA/cm^{-2}, 35 °C, 150 rpm; (**d**) SiC*w* 0.8 g L^{-1}, 18 mA/cm^2, 50 °C, 150 rpm; (**e**) SiC*w* 0.8 g L^{-1}, 18 mA/cm^2, 50 °C, 300 rpm.

Table 2. Weight fraction of the composite coatings as deposited at different conditions.

Samples	Weight Fraction (wt. %)			
	Ni	Si	C	O
a	98.66	—	0.32	1.02
b	95.18	1.47	0.55	2.80
c	90.88	3.48	1.00	4.64
d	85.57	5.39	1.48	5.56
e	76.32	8.83	4.67	10.18

3.2. Effects of Electrodeposition Parameters on Crystallographic Texture of the Ni/ SiCw Composite Coatings

For all the electrodeposited coatings, the nickel crystals are face-centered cubic structures (FCC). The three peaks at 2θ = 44.5, 51.85 and 76.37° that corresponds to (111), (200) and (220) crystallographic planes of nickel respectively, are in agreement with the standard XRD pattern JCPDS 04-0850. Note that the peaks corresponding to the SiC*w* were not detected in the XRD pattern of the composites. This might be due to the high content of Ni and its high scattering factor leading to higher peak intensities compared to those of SiC*w* [25]. Figure 6 shows the XRD pattern for the applied SiC whiskers. Comparing to the standard card, SiC particles was fit to No. 29-1129 exactly.

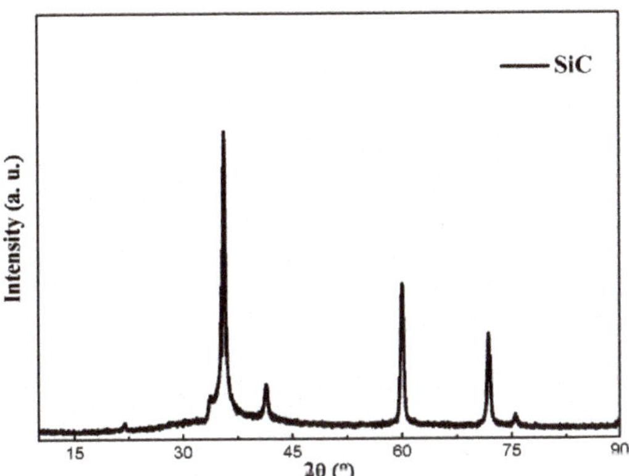

Figure 6. XRD pattern of the SiC whiskers.

Figure 7 shows the XRD patterns and its crystallographic orientation indexes of the Ni/SiC*w* composite coatings fabricated at a different current density. As shown in Figure 7a, at low current density (2 mA·cm^{-2}), the composite coating exhibits a strong (111) plane preferred orientation and a relatively low intensity of peaks (200) and (220). However, the intensity of (111) and (220) peak gradually reduce with the increase of current density, while the (200) peak increased obviously. Particularly, when the current density increased to 40 mA·cm^{-2}, the value of (200) peak intensity reaches the strongest, and the other peaks are almost disappeared. These results mean that the current density alters the texture of the composites and enhances the (200) reflections, which is in good agreement with the literature reported observation (the current efficiency of Ni deposition reached high values of over 97%) [26]. XRD analyses show that the texture coefficient of the deposits varies with current density. From Figure 7b and Table 3, the calculated results exhibit that the texture coefficient of (111) and (220) orientation decreases from 0.91 to 0.05 and 0.43 to 0.18, respectively; while the (200) orientation index increases from 0.66 to 2.9.

The variation of preferred orientation implied two possible reasons. One is the growth axis of the nickel grain mainly depends on the applied current density. There is a trend for dominating surface textures with increasing current density, as follows—(110) < (211) < (100). This means that (110), (211) and (100) texture are preferred orientation at very low current, medium and high current, respectively [27]. This also can be explained by the theory of nucleation energy (W_{hkl}) [28,29] as Equation (2).

$$W_{hkl} = \frac{B_{hkl}}{\frac{zF}{Na}\eta - A_{hkl}}, \quad (2)$$

where A_{hkl} and B_{hkl} are material parameters of the crystallographic direction (*hkl*); Na and F are the constant of Avogadro and Faraday, respectively; z is the number of electrons, and η is the over-voltage of cathodic polarization. For face-centered cubic metals, at low over-voltage, W111 < W100 < W110, while W100 < W111 at high over-voltage [26]. This phenomenon indicates that when particular nucleation energy is lower than other types of nucleation energy, the crystals of the deposited metal will exhibit particular preferred orientation. The other is the incorporated whiskers provides the occurrence of new nuclei and limits the growth of the original crystal grains [30,31], which facilitates more nucleation sites and the grain size refined. As mentioned above, the crystalline structure of Ni-W/SiC deposits is also affected by the factor of the SiC*w* content in the composites.

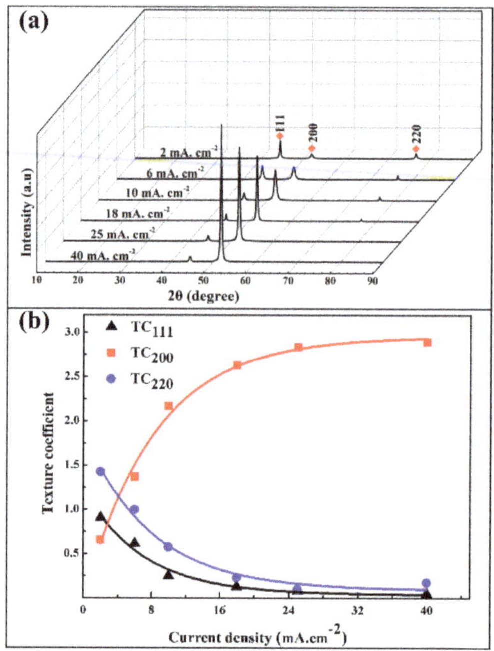

Figure 7. (**a**) XRD patterns of composites fabricated at a different current density, (**b**) The crystallographic orientation indexes of (111), (200) and (220).

Table 3. Texture coefficients of various (*hkl*) for the composite coatings as deposited at different current density.

(*hkl*)	$TC_{(hkl)}$					
	2 mA·cm^{-2}	6 mA·cm^{-2}	10 mA·cm^{-2}	18 mA·cm^{-2}	25 mA·cm^{-2}	40 mA·cm^{-2}
111	0.91	0.62	0.25	0.13	0.08	0.05
200	0.66	1.73	2.17	2.63	2.84	2.9
220	1.43	1	0.58	0.23	0.1	0.18

Figure 8 shows the XRD patterns and its crystallographic orientation indexes of the Ni/SiC*w* composite coatings electroplated at 20, 30, 40, 50, 60 °C, respectively. With the temperature elevated, the intensity of the (200) increases, while (111), (220) peaks increases. The calculated orientation index of (200) plan increases from 1.08 to 2.32, while the (111) and (220) orientation index decreases from 1.15 to 0.32 and 0.73 to 0.36 (Figure 8b and Table 4), respectively. These results indicate that relative high electrolyte temperature is benefit for the preferred orientation of (200) plan, which is in agreement with previous studies [32].

In the view of the observations made by E. Budevski et al. [28], the energy barrier of nucleation showed a positive linear dependence with 1/η and 1/η2 for the 2D and 3D model, respectively; the nucleation kinetics of nickel crystallites depends on the applied overpotential η. At low overvoltage, the particles near the cathode suppress metal ion reduction; while at high overvoltage, the transmission of the metal ions to the cathode becomes an important factor, moreover, the incorporated of inert whiskers enhance this transmission [23]. Therefore, the grains size of composites at high temperature becomes larger than that at low temperature (Figure 2).

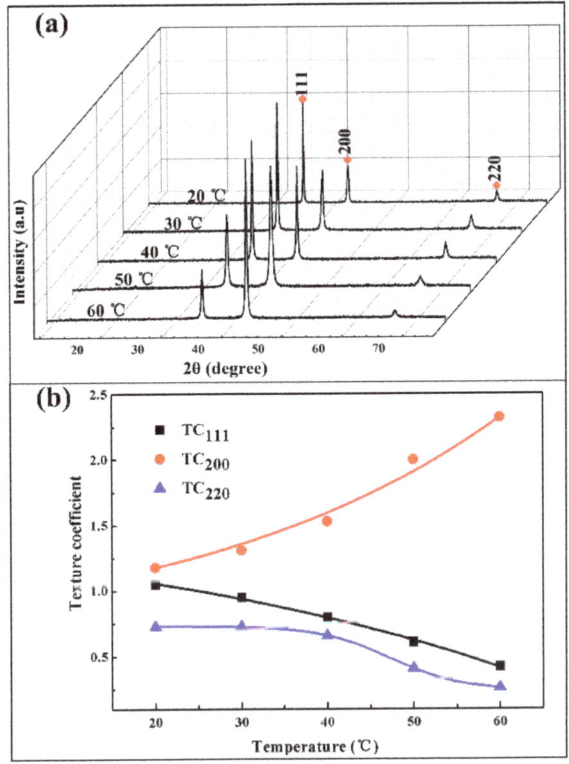

Figure 8. (**a**) XRD patterns of composites fabricated at a different temperature, (**b**) The crystallographic orientation indexes of (111), (200) and (220).

Table 4. Texture coefficients of various (*hkl*) for the composite coatings as deposited at different temperature.

(*hkl*)	$TC_{(hkl)}$				
	20 °C	30 °C	40 °C	50 °C	60 °C
111	1.15	1.05	0.8	0.51	0.32
200	1.08	1.21	1.53	2	2.32
220	0.73	0.73	0.71	0.51	0.36

Figure 9 shows the XRD patterns and its crystallographic orientation indexes of the Ni/SiC*w* composite coatings electroplated under different stirring conditions. The XRD patterns (Figure 9a) shows that, under static conditions, the intensity of (200) peak is slightly higher than that of (220) peak, and (111) peak possess the highest intensity. When given a certain stirring speed at 100 rpm, there is about the same intensity of (111) and (200) crystal planes, while the intensity of (220) peak is very weak and almost disappeared. With increasing stirring speed to 300 rpm, the intensity of (200) peaks gradually increase, along with a relative decrease in intensity of (111) and (220) peaks. Notably, at high stirring speed (500 rpm), the intensity of (200) peak presents a weakening trend. This is because the excessive stirring speed takes a large impact force for the cathode surface, it can, therefore, caused the whiskers transferred with the electrolyte at a high speed. Ultimately, the content of whiskers incorporated into the nickel matrix is reduced. The XRD patterns further analysis is shown in Figure 9b and Table 5. The preferred orientation of the composite coating at different stirring speeds is mainly related to the content of SiC*w* in coatings, and thereby the embedded SiC whiskers into the nickel matrix change the microstructure of the composites.

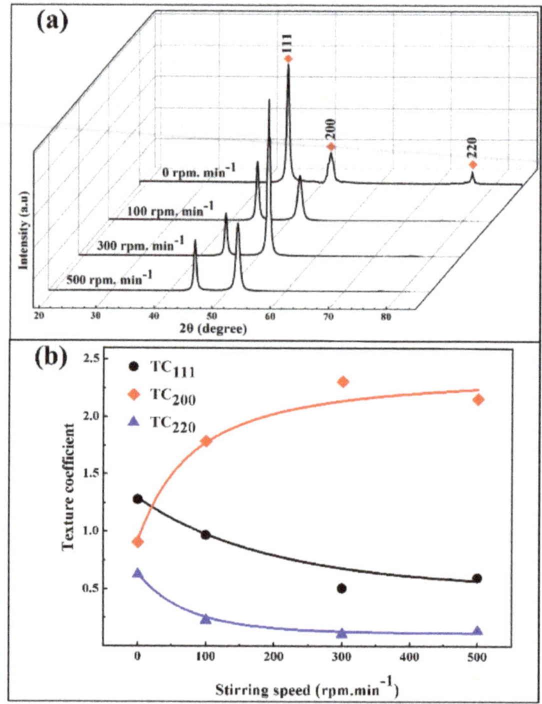

Figure 9. (**a**) XRD patterns of composites fabricated at a different stirring speed; (**b**) the crystallographic orientation indexes of (111), (200) and (220).

Table 5. Texture coefficients of various (*hkl*) for the composite coatings as deposited at different stirring speed.

(*hkl*)	$TC_{(hkl)}$			
	0 rpm	100 rpm	300 rpm	500 rpm
111	1.28	0.97	0.31	0.69
200	0.88	1.79	2.58	2.16
220	0.83	0.23	0.11	0.14

4. Conclusions

In summary, the influences of the external conditions of electrodeposition on the crystallographic texture, morphology and composition were investigated. For the deposited Ni/SiCw composite coatings, the content of SiC*w* in coatings increased with the increasing current density, stirring rate and electrolyte temperature in a certain range. Particularly, glossy Ni matrix surface with uniformly distributed whiskers was obtained at a current density of 18 mA/cm^2, whereas it became nodular and inhomogeneous with the widely differed spherical particles. Moreover, the detailed XRD analysis indicated that the current density altered the texture of the composites and enhanced the reflections of (200) plane. Also, the optimized temperature results revealed that the flat and bright morphology, as well as high content of whiskers in the composite coatings, was obtained at a temperature of 50 °C. Furthermore, the effects of stirring speed on the composition and structure on the coatings were studied. It was found that when the stirring speed was too low, the composite coatings appeared dendrites or nodules, while the whisker content was drastically reduced at a high speed. Consequently, these external conditions not only affect the surface morphology and the SiCw content of the composites, but also modified their preferred orientation.

Author Contributions: Conceptualization, G.D. and L.L.; methodology, L.L. and H.W.; software, L.L. and Y.S.; validation, G.D.; formal analysis, L.L. and H.W.; investigation, Z.Y. and Y.S.; resources, G.D.; data curation, L.L. and G.D; writing—original draft preparation, L.L.; writing—review and editing, L.L. and G.D.; visualization, H.W.; supervision, Y.Z.; project administration, G.D.; funding acquisition, G.D.

Funding: This research was funded and supported by the National Natural Science Foundation of China (No. 61571287).

Acknowledgments: The authors would like to thank supports from the Shanghai Professional Technical Service Platform for Non-Silicon Micro-Nano Integrated Manufacturing.

Conflicts of Interest: The authors declare no conflict of interest.

References

1. Li, B.; Zhang, W.; Zhang, W.; Huan, Y. Preparation of Ni-W/SiC nanocomposite coatings by electrochemical deposition. *J. Alloy. Compd.* **2017**, *702*, 38–50. [CrossRef]
2. Fink, C.G.; Gray, O.H. Co-Deposition of Lead and Bismuth. *Trans. Electrochem. Soc.* **1932**, *62*, 123. [CrossRef]
3. Grazen, A.E. Method for Electroforming and Coating. U.S. Patent No. 3,061,525, 30 October 1962.
4. Ekmekci, D.; Bülbül, F. Preparation and characterization of electroless Ni–B/nano-SiO_2, Al_2O_3, TiO_2 and CuO composite coatings. *Bull. Mater. Sci.* **2015**, *38*, 761–768. [CrossRef]
5. Madah, F.; Dehghanian, C.; Amadeh, A.A. Investigations on the wear mechanisms of electroless Ni–B coating during dry sliding and endurance life of the worn surfaces. *Surf. Coat. Technol.* **2015**, *282*, 6–15. [CrossRef]
6. Zhang, W.; Li, B. Influence of electrodeposition conditions on the microstructure and hardness of Ni-B/SiC nanocomposite coatings. *Int. J. Electrochem. Sci.* **2018**, *13*, 3486–3500. [CrossRef]
7. Moghadam, A.D.; Omrani, E.; Menezes, P.L.; Rohatgi, P.K. Mechanical and tribological properties of self-lubricating metal matrix nanocomposites reinforced by carbon nanotubes (CNTs) andgraphene—A review. *Compos. Part B* **2015**, *77*, 402–420. [CrossRef]
8. Snead, L.L.; Nozawa, T.; Katoh, Y.; Byun, T.S.; Kondo, S.; Petti, D.A. Handbook of SiC properties for fuel performance modeling. *J. Nucl. Mater.* **2007**, *371*, 329–377. [CrossRef]
9. Benea, L.; Bonora, P.L.; Borello, A.; Martelli, S. Wear corrosion properties of nano-structured SiC–nickel composite coatings obtained by electroplating. *Wear* **2001**, *249*, 995–1003. [CrossRef]
10. Benea, L.; Wenger, F.; Ponthiaux, P.; Celis, J.P. Tribocorrosion behaviour of Ni–SiC nano-structured composite coatings obtained by electrodeposition. *Wear* **2009**, *266*, 398–405. [CrossRef]
11. Lekka, M.; Kouloumbi, N.; Gajo, M.; Bonora, P.L. Corrosion and wear resistant electrodeposited composite coatings. *Electrochim. Acta* **2005**, *50*, 4551–4556. [CrossRef]
12. Ye, Z.; He, Q.; Lang, Y.; Chen, Y.; Liu, L.; Chen, C.; Ma, G. The effect of current density on the corrosion resistance and wear resistance of electrodeposition silver-Graphite composite coating. *J. Funct. Mater.* **2016**, *47*, 08227–08231.

13. Hai, J.C.; Tam, J.; Kovylina, M.; Kim, Y.J.; Erb, U. Thermal conductivity of bulk nanocrystalline nickel-diamond composites produced by electrodeposition. *J. Alloy. Compd.* **2016**, *687*, 570–578.
14. Jun, C.H.; Young, J.K.; Uwe, E. Thermal conductivity of copper-Diamond composite materials produced by electrodeposition and the effect of TiC coatings on diamond particles. *Compos. Part B* **2018**, *155*, 197–203.
15. Ueda, M.; Arai, S. Fabrication of High Thermal Conductivity Cu/Diamond Composites by Electrodeposition. *Electrochem. Soc.* **2014**, *49*, 2220.
16. Gül, H.; Kılıç, F.; Uysal, M.; Aslan, S.; Alp, A.; Akbulut, H. Effect of particle concentration on the structure and tribological properties of submicron particle SiC reinforced Ni metal matrix composite (MMC) coatings produced by electrodeposition. *Appl. Surf. Sci.* **2012**, *258*, 4260–4267. [CrossRef]
17. Gyftou, P.; Pavlatou, E.; Spyrellis, N. Effect of pulse electrodeposition parameters on the properties of Ni/nano-SiC composites. *Appl. Surf. Sci.* **2008**, *254*, 5910–5916. [CrossRef]
18. Chung, C.K.; Chang, W.; Chen, C.; Liao, M. Effect of temperature on the evolution of diffusivity, microstructure and hardness of nanocrystalline nickel films electrodeposited at low temperatures. *Mater. Lett.* **2011**, *65*, 416–419. [CrossRef]
19. Borkar, T.; Harimkar, S.P. Effect of electrodeposition conditions and reinforcement content on microstructure and tribological properties of nickel composite coatings. *Surf. Coat. Technol.* **2011**, *205*, 4124–4134. [CrossRef]
20. Chen, N.; Cheng, L.; Liu, Y.; Fang, Y.; Li, M.; Gao, Z.; Zhang, L. Microstructure and properties of SiC w /SiC composites prepared by gel-Casting combined with precursor infiltration and pyrolysis. *Ceram. Int.* **2018**, *44*, 969–979. [CrossRef]
21. Harimkar, S.P.; Dahotre, N.B. Crystallographic and morphological textures in laser surface modified alumina ceramic. *J. Appl. Phys.* **2006**, *100*, 189. [CrossRef]
22. Hashemzadeh, M.; Raeissi, K.; Ashrafizadeh, F.; Khorsand, S. Effect of ammonium chloride on microstructure, super-hydrophobicity and corrosion resistance of nickel coatings. *Surf. Coat. Technol.* **2015**, *283*, 318–328. [CrossRef]
23. Hovestad, A.; Janssen, L. Electrochemical codeposition of inert particles in a metallic matrix. *J. Appl. Electrochem.* **1995**, *25*, 519–527. [CrossRef]
24. Zhang, R.; Gao, L.; Guo, J. Preparation and characterization of coated nanoscale Cu/SiCp composite particles. *Ceram. Int.* **2004**, *30*, 401–404. [CrossRef]
25. Li, Y.; Wang, G.; Liu, S.; Zhao, S.; Zhang, K. The preparation of Ni/GO composite foils and the enhancement effects of GO in mechanical properties. *Compos. Part B* **2018**, *135*, 43–48. [CrossRef]
26. Burzyńska, L.; Rudnik, E.; Baz, L.; Kotula, M.; Sierpiński, Z.; Szymański, W. The influence of current density and bath composition on the electrodeposition of nickel and nickel/silicon carbide composite. *Trans. IMF* **2003**, *81*, 193–198. [CrossRef]
27. Amblard, J.; Epelboin, I.; Froment, M.; Maurin, G. Inhibition and nickel electrocrystallization. *J. Appl. Electrochem.* **1979**, *9*, 233–242. [CrossRef]
28. Budevski, E.; Staikov, G.; Lorenz, W. Electrocrystallization: Nucleation and growth phenomena. *Electrochim. Acta* **2000**, *45*, 2559–2574. [CrossRef]
29. Pangarov, N.; Velinov, V. The orientation of silver nuclei on a platinum substrate. *Electrochim. Acta* **1966**, *11*, 1753–1758. [CrossRef]
30. Momeni, M.; Hashemizadeh, S.; Mirhosseini, M.; Kazempour, A.; Hosseinizadeh, S. Preparation, characterisation, hardness and antibacterial properties of Zn–Ni–TiO2 nanocomposites coatings. *Surf. Eng.* **2016**, *32*, 490–494. [CrossRef]
31. Niksefat, V.; Ghorbani, M. Mechanical and electrochemical properties of ultrasonic-assisted electroless deposition of Ni–B–TiO2 composite coatings. *J. Alloy. Compd.* **2015**, *633*, 127–136. [CrossRef]
32. Wasekar, N.P.; Haridoss, P.; Seshadri, S.K.; Sundararajan, G. Influence of mode of electrodeposition, current density and saccharin on the microstructure and hardness of electrodeposited nanocrystalline nickel coatings. *Surf. Coat. Technol.* **2016**, *291*, 130–140. [CrossRef]

Article

Optimization of Processing Parameters for Water-Jet-Assisted Laser Etching of Polycrystalline Silicon

Xuehui Chen, Xiang Li, Chao Wu, Yuping Ma, Yao Zhang, Lei Huang * and Wei Liu

College of Mechanical and Electrical Engineering, Anhui Jianzhu University, Hefei 230601, China; xhenxh@163.com (X.C.); lix1117@163.com (X.L.); 15212428852@163.com (C.W.); wxlmyp@163.com (Y.M.); nianmin134@163.com (Y.Z.); wliu@hfcas.ac.cn (W.L.)

* Correspondence: huangl75@ahjzu.edu.cn; Tel.: +86-137-2110-2968

Received: 11 April 2019; Accepted: 1 May 2019; Published: 8 May 2019

Abstract: Liquid-assisted laser technology is used to etch defect-free materials for high-precision electronics and machinery. This study investigates water-jet-assisted laser etching of polysilicon material. The depths and widths of the etched grooves were investigated for different water-jet incident angles and velocities. To select optimal parameters for a composite etching processing, the results of many tests must be compared, and at least one set of good processing parameter combinations must be identified. Herein, the influence of different parameters on the processing results is studied using an orthogonal test method. The results demonstrate that the depths and widths of the processing grooves were nearly identical at water-jet angles of 30°and 60°; however, the 60° incidence conferred a slight advantage over 30° incidence. The section taper, section depth, and surface topography were optimized at a water-jet velocity of 24-m/s, 1.1-ms laser pulse width, 40-Hz frequency, and 180-A current. Under these conditions, the section taper and groove depth were 1.2° and 1.88 mm, respectively. The groove surfaces exhibited no splitting, slagging, or other defects, and no recast layers were visible.

Keywords: laser etching; water jet; polycrystalline silicon; orthogonal test

---

1. Introduction

Polysilicon material is a brittle semiconductor material with high wear resistance, hardness, chemical stability, and low thermal conductivity. The superior performance of silicon materials has secured their use in photovoltaics, aerospace, electronics, and other industries. However, owing to their physical properties, polysilicon and similar semiconductor materials are difficult to process effectively via conventional machining, i.e., the quality of processing may be poor. This problem can be resolved by employing laser processing technology. Such technology is widely used in electronic and other industries for precision machining of difficult-to-machine metal and non-metal material with high hardness, strength, toughness, and brittleness [1–3]. Unfortunately, under ambient air or inert gas conditions, the transient action of laser heat in the laser processing of materials inevitably produces microcracks and slag on the surface of the material, thereby reducing the qualification rate and processing quality of precision device products. To ameliorate the shortcomings of traditional lasers in processing such materials, liquid-assisted laser technology has been proposed herein. In this technology, with a hard and brittle material, the liquid reduces the temperature gradient at the processing site, thereby avoiding cracking and heat-affected areas of the material and improving the processing quality. Over time, liquid-assisted laser processing technology has branched into water-guided, underwater, and water-jet-assisted laser processing technology. Li et al. [4] etched trenches into silicon using a water-guided laser micromachining method. They simulated the melting and removal of silicon under the cooling effect of the water jet and verified their model via comparative simulations and experiments. Their model considers the effects of cutting speed and other factors on the surface quality and heat-affected zone of the workpiece. Porter et al. [5] applied water-guided laser processing technology to cut metal sheets and summarized the influence of some processing parameters on the results. Hock et al. [6] cut stainless-steel plates and brass sheets (thickness $\leq$ 100 μm) using conventional laser and water-guided laser cutting technologies and compared the cutting widths, heat-affected zones of the two products. Compared to the conventional technique, the water-guided laser cutting system achieved a lower heat-affected zone, slag accumulation height, and slit width. Adelmann et al. [7] cut aluminum, titanium, steel, and other metals using a water-guided laser and investigated the effects of laser power, pulse repetition frequency and etching times on the processing quality. The maximum depths of the cut aluminum, titanium, and steel were 8, 4.7, and 1.5 mm, respectively, and the maximum aspect ratios were 66.7, 39.2 and 12.5, respectively. Using Nd:YAG pulse fiber lasers, Choubey et al. [8] conducted stainless-steel plate experiments in both air and underwater environments. Their results demonstrate that compared to air cutting, underwater cutting improves the sample and slightly reduces its slit width, heat-affected zone, and adhesion slag. Mullick et al. [9] established a model for investigating the energy-loss mechanism in underwater laser cutting. The model accounts for the interaction between the laser, material, and water. Most of the energy is lost by the scattering of the laser light in water vapor. The losses account for 40–50% of the total laser energy. The correctness of the model results was verified via experiments. Tsai [10] processed LCD glass materials using underwater and conventional laser techniques and investigated the effects of the laser parameters, material size, and liquid type on the removal rate, slit width, incision depth, taper, and surface roughness of the processed materials. Their results demonstrated the advantages of underwater laser processing over conventional laser processing. Charee et al. [11] processed silicon wafers via a flowing underwater laser composite etching process. To study the effects of underwater laser etching, they controlled the rate and direction of the water flow in a closed water chamber. With a high water flow rate, the grooves were deepened and few re-agglomeration layers were formed. The etching effect was superior to that obtained in still water. Kalyanasundaram [12] combined the parameters of laser water jets with those of hard and brittle processing materials and significantly improved the quality of the cut. Tangwarodomnukun et al. [13,14] compared the surface morphologies of samples prepared via hybrid laser–water jet micromachining, traditional silicon processing, and laser composite processing technologies. They varied the processing parameters and analyzed their effect on the heat-affected zone and processing quality. Bao et al. [15] established a fluid dynamics model based on smoothed particle hydrodynamics experimentally investigated the water-jet-assisted laser cutting of silicon materials. Experimental analysis revealed that the laser ablation of silicon mainly occurs via explosive melting. Zhu et al. [16] established a numerical model of heat transfer and material ablation in the water-jet-assisted laser etching of single-crystal germanium. They reported that the irradiated material can be discharged through the water jet and that during the non-pulsing periods, the water-jet cooling effectively removes heat buildup in the workpiece, thereby minimizing the thermal damage caused by laser heating. They also analyzed the influence of the processing parameters on the ablation process. Increasing the laser pulse energy deepened the grooves, whereas increasing the applied water pressure reduced the threshold workpiece temperature for material removal. Feng et al. [17] established a three-dimensional analytical model of the temperature field during the hybrid laser water-jet micromachining process. Their model considers the interaction between the laser, water jet, and workpiece. The absorption of laser light by the water, the formation of laser-induced plasma in water, the formation of bubbles, and the laser refraction at the air–water interface were discussed. To evaluate two machining methods, Hao Zhu et al. [18] compared the direct and chemical-assisted picosecond laser trepanning of single crystalline silicon. In their study, an orthogonal test design scheme was adopted to consider the relevant parameters affecting the trepanning process. They found that the direct laser trepanning results are associated with significant thermal defects, whereas the chemical-assisted method can process microholes with negligible thermal damage.

Conventional gas-assisted laser processed materials have defects such as a large heat-affected zone and many re-casting layers. Compared with traditional gas-assisted laser processing, the laser processing area is cooled and impacted by the water jet. Specifically, during laser processing, the water jet cools the material and its impact can wash away the slag and recast a layerafter laser processing. The cooling action can reduce excess heat acting on the surface of the material, reduce the heat-affected zone, and reduce the generation of microcracks, thereby improving the surface quality of the material. In this paper, water jets with different angle and velocities were used to assist laser etching of polysilicon. Previously, we studied the effect of different water jet velocities on the processing grooves when the water jet angle was 30° [19]. However, to the best of our knowledge, the effects of changing both the water-jet angle and velocity on the depth and width of the etched grooves have not been investigated. This paper builds on previous studies by varying the jet angles and velocities in water-jet-assisted laser etching of polysilicon materials and examining its effect on the depth and width of the etched grooves. Subsequently, after fixing the water-jet angle at 60°, the optimal processing parameters in the experimental grooves are investigated in an orthogonal optimization experiment.

2. Theoretical Analysis of the Influence of Water-Jet-Assisted Laser Etching

Water-jet-assisted laser processing introduces a water jet to traditional laser processing. The auxiliary gas removes slag and debris generated in the laser processing, thereby improving the processing quality of the materials and reducing slag accumulation, the generation of microcracks, formation of recast layers during processing, and the heat-affected zone. Figure 1 schematizes the water-jet-assisted laser technique. Note that the water jet consumes part of the laser energy. In contrast, when jetted at appropriate velocity, the water jet washes away slag and the recast layer in an appropriate time, which reduces the heat-affected zone and improves processing quality. Water-jet-assisted laser processing also reduces photothermal efficiency and suppresses secondary adhesion of the etched material via laser-induced liquid cavitation, water cooling, and water flow.

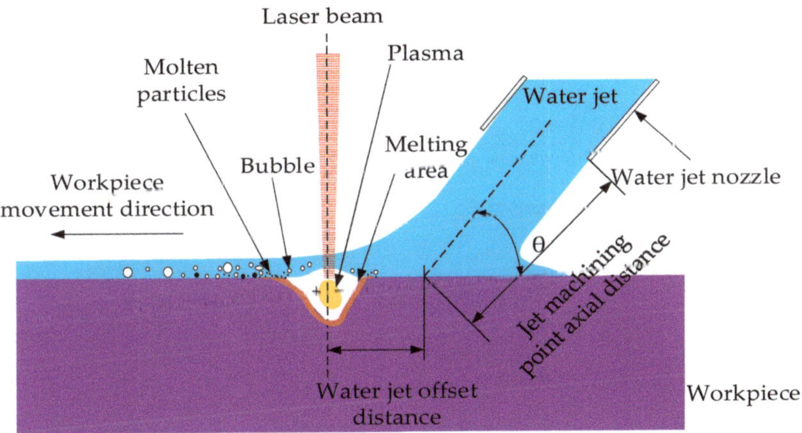

Figure 1. Schematic of water-jet-assisted laser processing.

Conventional gas-assisted and water-jet-assisted laser processing techniques remove material via laser irradiation, which smelts or vaporizes the material surface. In the conventional technique, most laser energy is absorbed by the material. The material's thermal properties will transfer excess heat around the laser action zone, which enlarges the heat-affected zone and increases slag accumulation. In water-jet-assisted laser etching technology, conventional laser etching is supplemented by water jets applied at a certain angle and velocity. This technology combines the high efficiency of traditional gas-assisted laser processing with the advantages of water jets. The laser primarily acts as a softening material that melts or vaporizes the material as it is heated. The water jet strikes the laser processing area and cools it by heat transfer. Generally, the specific heat capacity of water is greater than that of the material matrix, which implies that water absorbs excess heat from the material surface and provides a cooling effect. In addition, increasing the water-jet velocity increases the impact force on the material. Within a certain range, the impact force is sufficient to flush the slag and other defects generated by the laser action. Then, the flowing water removes the slag or recast layer from the tank in sufficient time. Note that reducing slag and the recast layer from the bottom and walls of the trench can improve processing quality [20].

In this study, the effects of water jet pressure at different angles and velocities on the material were compared in fluent simulations. Herein, the water-jet angle was set to 30° or 60°, and the velocity was varied between 16, 20, 24, and 28 m/s (Figure 2). Note that many factors were simplified during simulation; thus, the results are theoretical rather than practical. As shown in Figure 2, increasing jet velocity at a fixed incident angle gradually increased the impact force of the water jet on the material. In addition, for a fixed water-jet velocity, the impact force was ~1.5 times higher at an impact angle of 60° than at 30°. Thus, it is theoretically demonstrated that 60° water-jet-assisted laser etching of the polysilicon material demonstrates groove depth and width greater than that of the 30° water-jet-assisted laser etching polysilicon material. In actual machining, greater jet velocity is advantageous for material removal; however, if the water flow velocity is excessive, the water jet will shake the experimental device, which will also deflect the impact center of the jet on the material. A significantly diverted water beam will cause splashing, which will generate "water mist" that affects the focused laser spot and incurs large laser energy loss.

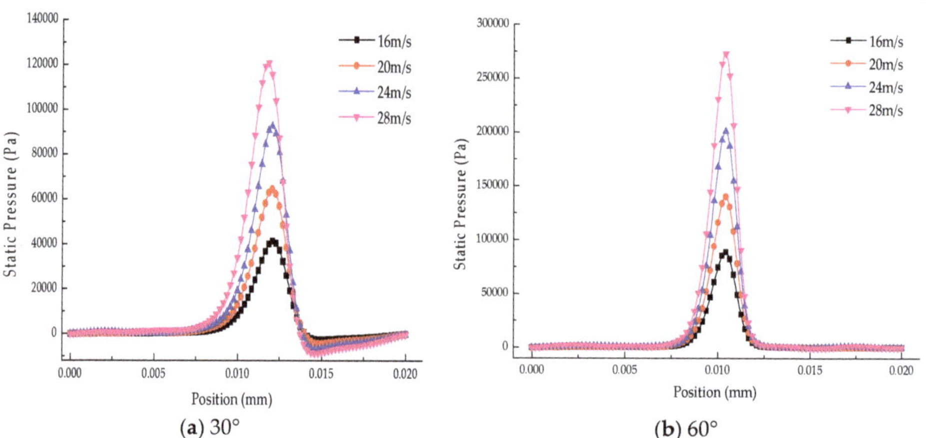

Figure 2. Impact force of a water jet striking a material at different velocities and impact angles.

3. Experimental Methods and Analysis

An HGL-LMY500 solid laser cutting and welding processing system (Wuhan Huagong Laser Engineering Co., Ltd., Wuhan, China) was used in the test. The HGL-LMY500 comprises an Nd:YAG solid-state pulsed laser (average output: 500 W), a power supply system, an optical system, a control system, a cooling system, a numerical control table, and a computer (Figure 3). During processing, this system produces a laser with a wavelength of 1064 nm, a multimode circular spot, and a spot diameter of 0.2 mm. By adjusting the current, pulse width, and repetition frequency, the laser output energy, laser light intensity, beam quality, etc. satisfy processing requirements. The main parameters of this laser system are shown in Table 1. Here, the water jet device is a water jet processing system designed by our team. Note that the water-jet velocity and angle be adjusted. In addition, the water jet nozzle (water jet nozzle diameter is 0.7 mm) can be fixed to ensure that the water jet can impact and cool the material in the laser processing area. According to the actual work requirement, the motor, liquid tank, filter device, inlet pipe, plunger pump, overflow return control valve, pressure gauge, accumulator, one-way control valve, and jet nozzle and its fixing device.

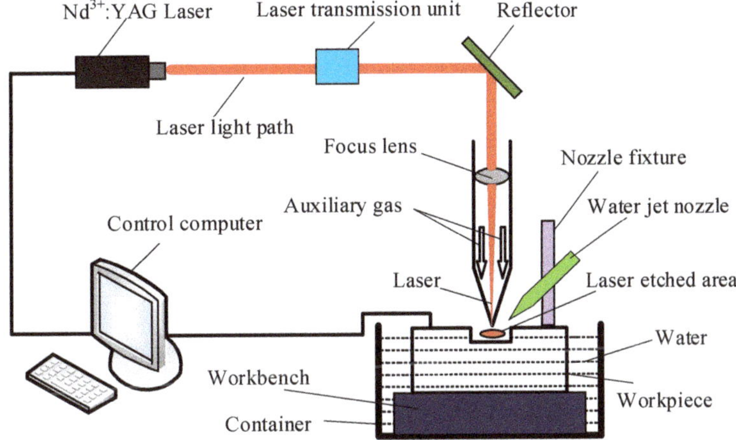

Figure 3. Schematic diagram of the experimental Nd^{3+}: YAG laser etching apparatus.

Table 1. Laser system processing parameters.

Technical Parameters	Current (A)	Pulse Width (ms)	Repeat Frequency (Hz)	Single Pulse Energy (J)	Processing Speed ($mm \cdot s^{-1}$)
Adjustment range	100–400	0.1–20	0–150	0–90	>0.1

The experimental sample was a polycrystalline silicon plate (length × width × height: 20 × 20 × 2 mm^3). The effect of different water jet angles and velocities on the depth and widths of the processing grooves in water-jet-assisted laser etching was investigated. Prior to experiment, it is necessary to ensure that the following laser parameters remain unchanged: 40-Hz laser repetition frequency, 0.6-ms pulse width, 1-mm/s scanning speed, and 140-A current. Under this processing parameter, the laser irradiation intensity reaches 5.5 × 10^5 W/cm^2. When the irradiation intensity is 5.5 × 10^5 W/cm^2, it is sufficient to achieve the energy density required for irreversible damage to the surface material, and the material is removed by the action of the laser. In water-jet-assisted laser processing, the jet is impacted slightly behind the laser beam (1 mm behind processing area) to avoid direct contact between the jet and laser beam, avoid excessive loss of laser energy, and boost cooling and impact performance. The cross-sections and widths of the resulting grooves are shown in Figures 4 and 5, respectively. Figures 6 and 7 show the groove depths and widths under different water-jet velocity (horizontal axis) and impact angle (colors), respectively.

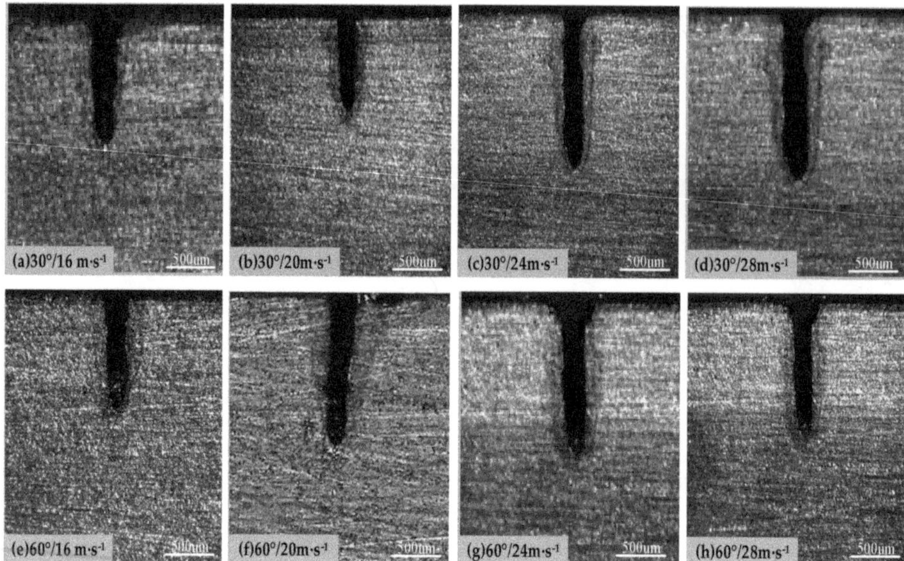

Figure 4. Cross-sections of grooves obtained by etching polycrystalline silicon at different impact angles (top row: 30°; bottom row: 60°) and jet velocities (left to right: 16, 20, 24, and 28 m/s) (25×).

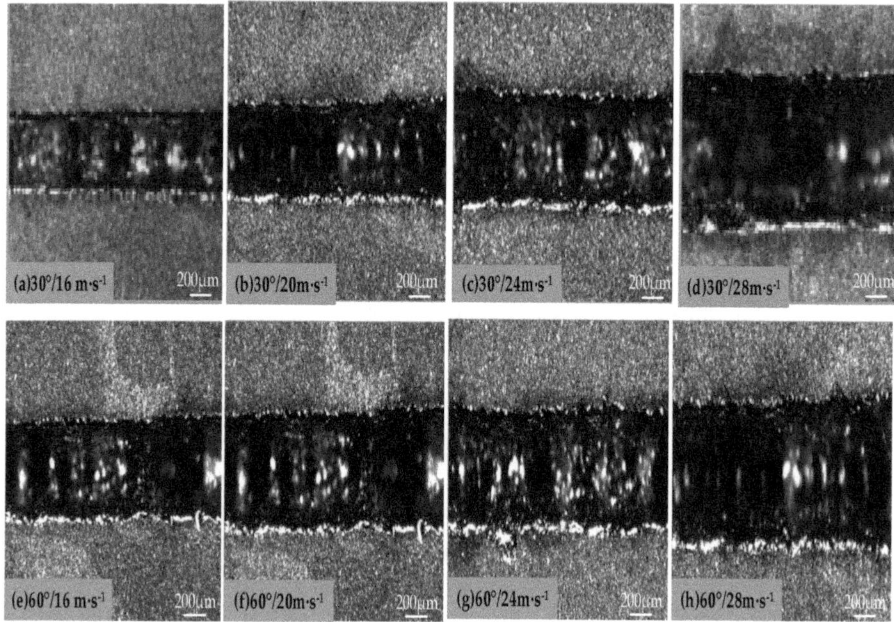

Figure 5. Widths of grooves obtained by laser etching at different impact angles (top row: 30°; bottom row: 60°) and jet velocities (left to right: 16, 20, 24, and 28 m/s) (30×).

enhanced convective heat transfer between the laser thermal energy and surrounding medium (i.e., the water) incurs large laser energy loss, which reduces the processing depth.

4. Orthogonal Test Design

4.1. Design Table of Orthogonal Test Plan

The orthogonal test method was employed to assess the quality of the water-jet-assisted laser etching of the polysilicon samples. In this experiment, the processing quality was primarily evaluated by examining the section taper, section morphology, and processing depth. At the largest processing depth, the processing tank must have the smallest taper and fine surface morphology. The cross-section of the sample trough was observed using a VM-3020E 2D imager, and the taper and depth of the section were measured. The results are shown in Figure 8.

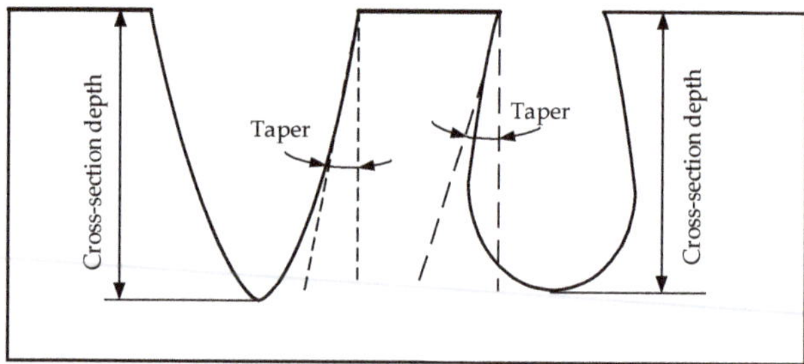

Figure 8. Cross-sectional profile of the groove in the polysilicon sample after processing.

Section 3 discussed the effect of water-jet angles and velocities on the depth and width of the grooves fabricated by laser-assisted etching of polysilicon material. Note that many interacting processing parameters are involved in the actual processing. The parameters that affect water-jet-assisted laser etching are the laser and water-jet parameters. To improve the etching effect, one or several sets of good combinations of processing parameters must be determined, which requires numerous experiments. The orthogonal test method optimizes the selection of processing parameters, which reduces the number of tests and the time required to select suitable processing parameters. As shown in Figure 2, the impact effect of the water jet at any incident velocity was slightly higher at 60° compared to that at 30°. Therefore, in the orthogonal test, the incident angle of the water jet was set to 60° and the other parameters were set according to those of the actual situation. Here, the main research objects were the laser processing parameters (laser pulse width, repetition frequency, and pulse energy) and pulse energy (represented by input current).

In the orthogonal test, the water-jet velocity, laser pulse width, laser repetition frequency, and laser input current were labeled *A*, *B*, *C*, and *D*, respectively, and each factor was assigned four levels (1, 2, 3, and 4), as shown in Table 2.

Table 2. Factor/level table.

Level	Factor			
	A Water-Jet Velocity/m/s	*B* Pulse Width/ms	*C* Repeat Frequency/Hz	*D* Current/A
1	16	0.5	30	150
2	20	0.7	35	160
3	24	0.9	40	170
4	28	1.1	45	180

The water-jet-assisted laser etching process was tested according to the L16 (45) orthogonal table (presented as Table 3). The design included 16 orthogonal experimental schedules. The subscripts in each plan indicate the level number (first column in Table 1). For example, A1 represents jet velocity of 16 m/s and B_4 indicates a pulse width of 1.1 ms.

Table 3. Orthogonal test table (L16 (45)).

Test Number	*A*	*B*	*C*	*D*	Test Plan
1	1	1	1	1	$A_1B_1C_1D_1$
2	1	2	2	2	$A_1B_2C_2D_2$
3	1	3	3	3	$A_1B_3C_3D_3$
4	1	4	4	4	$A_1B_4C_4D_4$
5	2	1	2	3	$A_2B_1C_2D_3$
6	2	2	1	4	$A_2B_2C_1D_4$
7	2	3	4	1	$A_2B_3C_4D_1$
8	2	4	3	2	$A_2B_4C_3D_2$
9	3	1	3	4	$A_3B_1C_3D_4$
10	3	2	4	3	$A_3B_2C_4D_3$
11	3	3	1	2	$A_3B_3C_1D_2$
12	3	4	2	1	$A_3B_4C_2D_1$
13	4	1	4	2	$A_4B_1C_4D_2$
14	4	2	3	1	$A_4B_2C_3D_1$
15	4	3	2	4	$A_4B_3C_2D_4$
16	4	4	1	3	$A_4B_4C_1D_3$

4.2. Experimental Results and Optimization Options

Following the test scheme shown in Table 3, an orthogonal test was conducted on the water-jet-assisted laser etching processing platform. Here, the measured data were the depth and taper of the groove of the given evaluation index. The 16 test results and their corresponding data are shown in Figures 9 and 10, respectively.

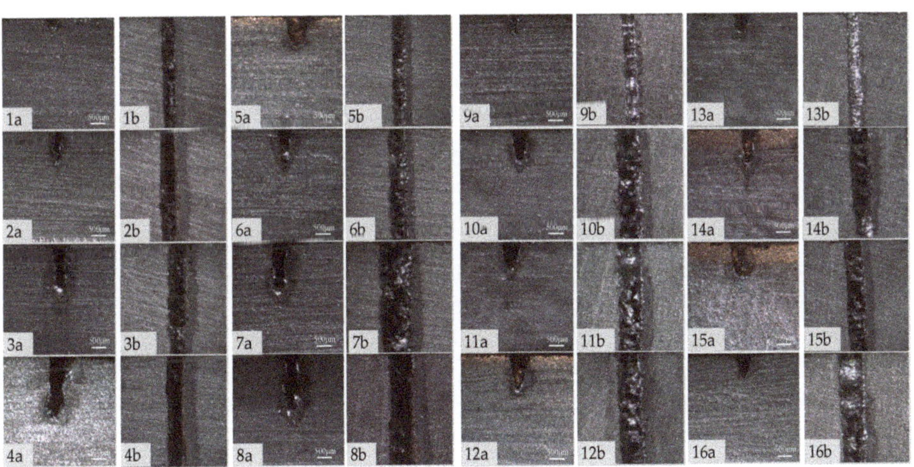

Figure 9. Orthogonal test results (15×). Label "a" represents the depth and taper of the section, and "b" represents surface topography.

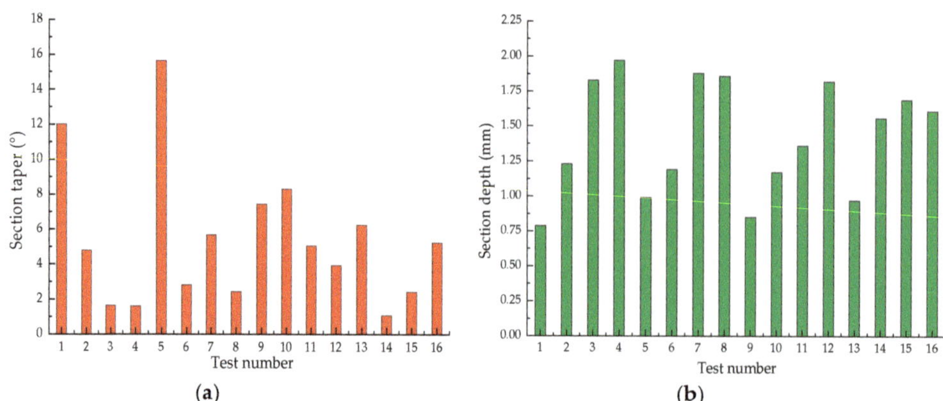

Figure 10. Experimental values of orthogonal tests 1–16. (**a**) Taper value; (**b**) Depth value.

The test results for each evaluation index in Table 4 were calculated and analyzed using the range analysis method. After computing test indices (i.e., section taper and section depth, corresponding to the m-th level of the j-th factor) and their average values (K_{jm}), the range differences R_j of the two evaluation indexes were calculated for the j-th factor. The results are shown in Table 4.

Table 4. Range calculation values of orthogonal test results.

Index		A	B	C	D
Section taper/°	$\overline{K}_{j1}$	5.025	10.342	6.285	5.682
	$\overline{K}_{j2}$	6.647	4.258	6.707	4.640
	$\overline{K}_{j3}$	6.200	3.710	3.162	7.717
	$\overline{K}_{j4}$	3.752	3.315	5.470	3.585
	R_j	2.895	7.027	3.545	4.132
Section depth/mm	$\overline{K}_{j1}$	1.455	0.900	1.238	1.513
	$\overline{K}_{j2}$	1.480	1.288	1.433	1.355
	$\overline{K}_{j3}$	1.300	1.690	1.525	1.400
	$\overline{K}_{j4}$	1.458	1.815	1.497	1.425
	R_j	0.180	0.915	0.287	0.158

From the average values of the indicators corresponding to each level of each factor, the appropriate level of each factor can be summarized as follows:

- If the required index must be as small as possible, we must consider the level corresponding to the smallest average.
- If the required index must be as large as possible, we must consider the level corresponding to the largest average.
- If the indicator must be moderate (a fixed value), we must consider the level corresponding to a moderate average.

To analyze the test results visually and intuitively and to obtain comprehensive conclusions, the relationships between the evaluation indicators and various factors (effect curves) were derived from the range analysis data. In this experiment, the average corresponding experimental indicator was plotted for each level in a given factor. The results of factors A, B, C, and D are shown in Figures 11–14, respectively.

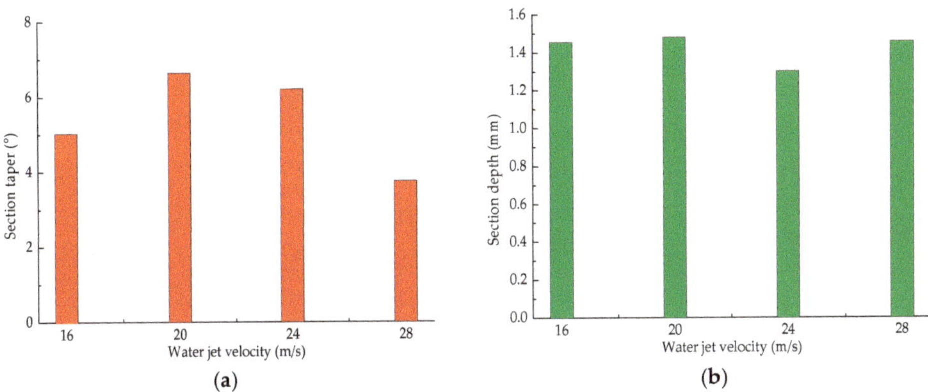

Figure 11. Effect of water-jet velocity on (**a**) cross-section taper and (**b**) cross-section depth.

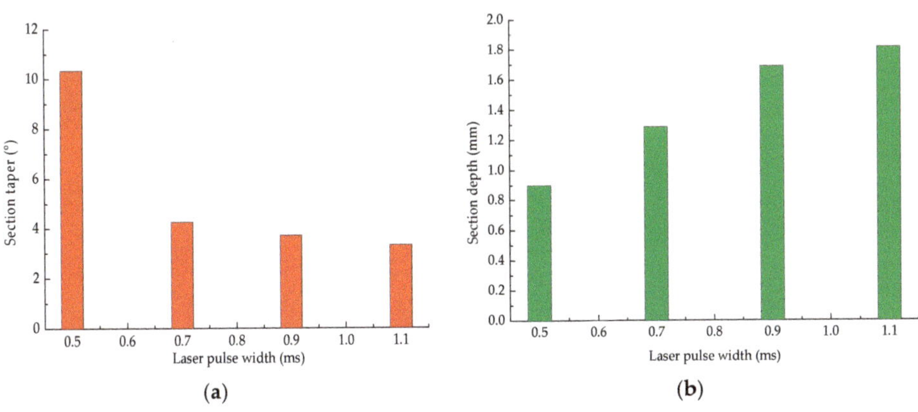

Figure 12. Effect of laser pulse width on (**a**) cross-section taper and (**b**) cross-section depth.

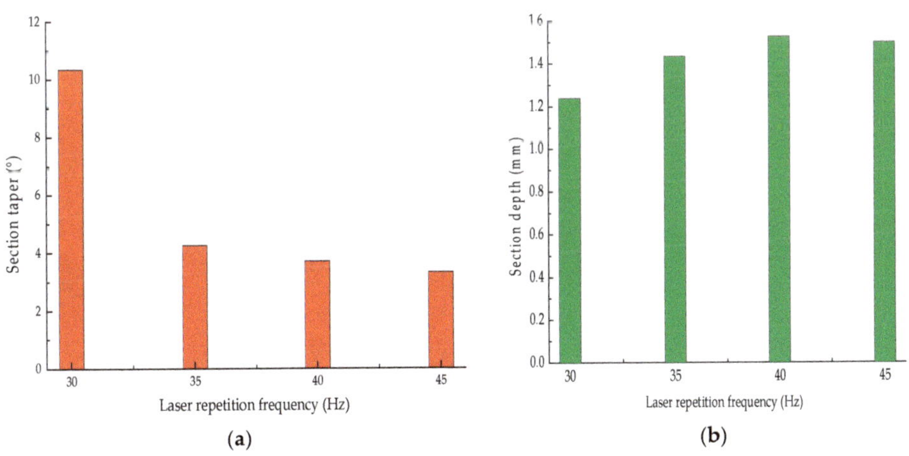

Figure 13. Effect of laser repetition frequency on (**a**) cross-section taper and (**b**) cross-section depth.

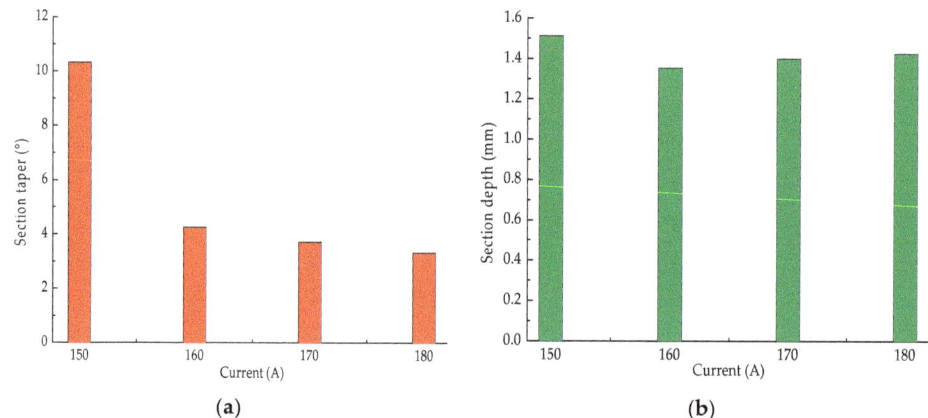

Figure 14. Effect of current on (**a**) cross-section taper and (**b**) cross-section depth.

The range analysis results given in Table 5 demonstrate that the section taper was most influenced by factor *B*, followed by factors *D*, *C*, and *A*. In addition, the actual processing requires that a smaller section taper will result in a larger section depth.

Table 5. Range analysis results of orthogonal test.

	Section Taper	Section Depth
Primary and secondary factors	B, D, C, A	B, C, A, D
Excellent level	A_4, B_4, C_3, D_4	A_2, B_4, C_3, D_1
Optimal combination	$B_4D_4C_3A_4$	$B_4C_3A_2D_1$

The primary and secondary factors, optimal level, and optimal combination (i.e., the optimal solution) were obtained (Table 6) by combining the range calculation results given in Table 4 with the relationships between evaluation indicators and factors shown in Figures 11–14.

Table 6. Processing parameters in optimized schemes of each evaluation index.

Evaluation Index	Actual Requirements	Excellent Solution	Water-Jet Velocity/m/s	Pulse Width/ms	Repeat Frequency/Hz	Current/A
Section taper/°	Smaller and better	$B_4D_4C_3A_4$	28	1.1	40	180
Section depth/mm	Bigger and better	$B_4C_3A_2D_1$	20	1.1	40	150

As shown in Table 6, the optimal process that minimizes the cross-section taper combines a 1.1-ms laser pulse width, 40-Hz frequency, and 180-A current with the 28-m/s water-jet velocity. Under this condition, the cross-section taper after composite processing was 0.3° (Figure 15a). The optimum process that maximizes the tank depth combines a 1.1-ms laser pulse width, 40-Hz frequency, and 150-A input current with the 20-m/s water-jet velocity. With this configuration, the etched tank had a depth of 2.05 mm but was shaped like a teardrop with a non-ideal taper (Figure 15b).

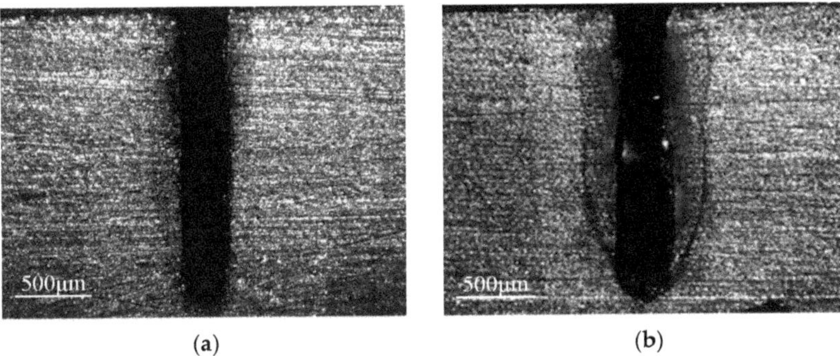

Figure 15. Cross-sectional morphology of samples after composite processing at processing parameters that optimize each evaluation index (×30): (**a**) section taper and (**b**) section depth.

Based on the above analysis, the final combination of optimal processing parameters was determined to be $B_4A_3C_3D_4$. The combination of processing parameters that optimized the cross-section taper and cross-section depth of the etched tank were as follows: jet velocity of 24 m/s, laser pulse width of 1.1 m/s, laser frequency of 40 Hz, and input current of 180 A. The cross-section and surface topography of the groove after composite processing under these conditions are shown in Figure 16. As can be seen, the surface of the groove body is free of splitting and slagging defects and has no recast layer. Here, the taper and depth of the groove were 1.2° and 1.88 mm, respectively. Although the etching result of each index was slightly worse than the result of optimizing a single indicator, it generally satisfies the etching effect requirements.

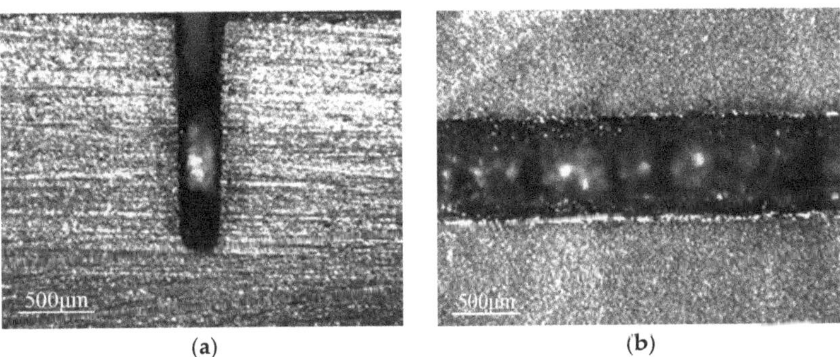

Figure 16. Surface and section morphologies of groove formed by composite machining under final optimized combination of processing parameters (×30): (**a**) sectional topography and (**b**) surface topography.

5. Conclusions

A water jet is frequently employed to assist laser etching of polysilicon materials. This study assessed the influence of water-jet incident angle and velocity on the depth and width of the etching groove. At a water jet incident at 60°, the effects of processing parameters (i.e., water-jet velocity, laser pulse width, repetition frequency, and input current) on the surface quality of the composite etching were investigated using the orthogonal test method. The processing parameters were optimized and verified. By comparing the verification test results with those of the orthogonal test, the processing parameters that optimize each evaluation index were obtained. Our primary findings are summarized as follows.

(1) The effects of 30° and 60° water-jet-assisted laser etching of polysilicon materials on the depth and width of the etched trenches were investigated. It was found that when the water jet angle was

maintained at one value, the depth of the machining tank initially increased and then decreased as water-jet velocity increased, and the width of the machining tank increased gradually. As the velocity increased from 16 to 24 m/s, the depth of the processed groove increased, but a further velocity increase (28 m/s) reduced the groove depth. The width of the processing groove was a gradually increasing function of the water-jet velocity. When the water-jet velocity was maintained at one value, the 60° water jet assisted the laser etching of the polysilicon material to obtain a tank with depth and width greater than that obtained by the 30° etching.

(2) Taking the 60° water-jet-assisted laser etching of polysilicon as an example, the orthogonal experimental method was used to optimize the processing parameters. In this study, the processing quality was evaluated relative to the depth, taper, and surface topography of the processing tank (the deeper the depth of the tank, the smaller the taper and the better the surface topography).

The experimental results demonstrated that the section taper was minimized under a 1.1-ms laser pulse width, 40-Hz frequency, and 180-A current when assisted by 28-m/s water-jet velocity. Here, the minimum section taper was 0.3°. The tank depth was maximized with a 1.1-ms laser pulse width, 40-Hz repetition frequency, and 150-A input current when assisted by a 20-m/s water-jet velocity. Here, the maximum tank depth was 2.05 mm; however, the section taper was not ideal. The optimal processing parameters were a laser pulse width of 1.1ms, repetition frequency of 40 Hz, water-jet velocity of 24 m/s, and input current of 180 A. Under these conditions, the cross-section taper and groove depth were 1.2° and 1.88 mm, respectively, and the groove surface demonstrated no defects.

Author Contributions: X.C. designed the study. Y.M., Y.Z., C.W., W.L. and L.H. developed the methodology. X.L. wrote manuscript.

Funding: This research was funded by Provincial Key Project of Natural Science Research Anhui Colleges (grant number: KJ2015A013 and KJ2015A050), Anhui Province Outstanding Young Talent Support Program Key Project (grant number: xyqZD2016153), National Natural Science Foundation of China (grant number: 51175229).

Acknowledgments: We thank all the authors for their joint efforts to complete the experiment. We also thank Jiangnan University for providing experimental equipment.

Conflicts of Interest: The authors declare no conflicts of interest.

References

1. Manickam, S.; Wang, J.; Huang, C. Laser–material interaction and grooving performance in ultrafast laser ablation of crystalline germanium under ambient conditions. *Proc. Inst. Mech. Eng. Part B J. Eng. Manuf.* **2013**, *227*, 1714–1723. [CrossRef]
2. Wu, Q.; Wang, J.; Huang, C. Analysis of the machining performance and surface integrity in laser milling of polycrystalline diamonds. *Proc. Inst. Mech. Eng. Part B J. Eng. Manuf.* **2014**, *228*, 903–917. [CrossRef]
3. Wang, J. An Experimental Analysis and Optimisation of the CO_2 Laser Cutting Process for Metallic Coated Sheet Steels. *Int. J. Adv. Manuf. Technol.* **2000**, *16*, 334–340. [CrossRef]
4. Li, C.F.; Johnson, D.B.; Kovacevic, R. Modeling of waterjet guided laser grooving of silicon. *Int. J. Mach. Tools Manuf.* **2003**, *43*, 925–936. [CrossRef]
5. Porter, J.A.; Louhisalmi, Y.A.; Karjalainen, J.A.; Füger, S. Cutting thin sheet metal with a water jet guided laser using various cutting distances, feed speeds and angles of incidence. *Int. J. Adv. Manuf. Technol.* **2007**, *33*, 961–967. [CrossRef]
6. Hock, K.; Adelmann, B.; Hellmann, R. Comparative study of remote fiber laser and water-jet guided laser cutting of thin metal sheets. *Phys. Procedia* **2012**, *39*, 225–231. [CrossRef]
7. Adelmann, B.; Ngo, C.; Hellmann, R. High aspect ratio cutting of metals using water jet guided laser. *Int. J. Adv. Manuf. Technol.* **2015**, *80*, 2053–2060. [CrossRef]
8. Choubey, A.; Jain, R.K.; Ali, S.; Singh, R.; Vishwakarma, S.C.; Agrawal, D.K.; Arya, R.; Kaul, R.; Upadhyaya, B.N.; Oak, S.M. Studies on pulsed nd:yag laser cutting of thick stainless steel in dry air and underwater environment for dismantling applications. *Opt. Laser Technol.* **2015**, *71*, 6–15. [CrossRef]
9. Mullick, S.; Madhukar, Y.K.; Roy, S.; Nath, A.K. An investigation of energy loss mechanisms in water-jet assisted underwater laser cutting process using an analytical model. *Int. J. Mach. Tools Manuf.* **2015**, *91*, 62–75. [CrossRef]

10. Tsai, C.H.; Li, C.C. Investigation of underwater laser drilling for brittle substrates. *J. Mater. Process. Technol.* **2009**, *209*, 2838–2846. [CrossRef]
11. Charee, W.; Tangwarodomnukun, V.; Dumkum, C. Laser ablation of silicon in water under different flow rates. *Int. J. Adv. Manuf. Technol.* **2015**, *78*, 19–29. [CrossRef]
12. Kalyanasundaram, D. Mechanics Guided Design of Hybrid Laser/Waterjet System for Machining Hard and Brittle Materials. Ph.D. Thesis, Iowa State University, Ames, IA, USA, 2009.
13. Tangwarodomnukun, V.; Wang, J.; Huang, C.Z.; Zhu, H.T. An investigation of hybrid laser-waterjet ablation of silicon substrates. *Int. J. Mach. Tools Manuf.* **2012**, *56*, 39–49. [CrossRef]
14. Tangwarodomnukun, V.; Wang, J.; Huang, C.Z.; Zhu, H.T. Heating and material removal process in hybrid laser-waterjet ablation of silicon substrates. *Int. J. Mach. Tools Manuf.* **2014**, *79*, 1–16. [CrossRef]
15. Bao, J.; Long, Y.; Tong, Y.; Yang, X.; Zhang, B.; Zhou, Z. Experiment and simulation study of laser dicing silicon with water-jet. *Appl. Surf. Sci.* **2016**, *387*, 491–496. [CrossRef]
16. Zhu, H.; Wang, J.; Yao, P.; Huang, C. Heat transfer and material ablation in hybrid laser-waterjet microgrooving of single crystalline germanium. *Int. J. Mach. Tools Manuf.* **2017**, *116*, 25–39. [CrossRef]
17. Feng, S.; Huang, C.; Wang, J.; Zhu, H.; Yao, P.; Liu, Z. An analytical model for the prediction of temperature distribution and evolution in hybrid laser-waterjet micro-machining. *Precis. Eng.* **2017**, *47*, 33–45. [CrossRef]
18. Zhu, H.; Zhang, Z.; Xu, K.; Xu, J.; Zhu, S.; Wang, A.; Qi, H. Performance Evaluation and Comparison between Direct and Chemical-Assisted Picosecond Laser Micro-Trepanning of Single Crystalline Silicon. *Materials* **2019**, *12*, 41. [CrossRef] [PubMed]
19. Chen, X.H.; Li, X.; Song, W.; Wu, C.; Zhang, Y. Effects of a low-pressure water jet assisting the laser etching of polycrystalline silicon. *Appl. Phys. A* **2018**, *124*, 556. [CrossRef]
20. Wang, L.; Huang, C.; Wang, J.; Zhu, H.; Liang, X. An experimental investigation on laser assisted waterjet micro-milling of silicon nitride ceramics. *Ceram. Int.* **2018**, *44*, 5636–5645. [CrossRef]

Article

Multi-Objective Design Optimization of the Reinforced Composite Roof in a Solar Vehicle

Ana Pavlovic, Davide Sintoni, Cristiano Fragassa * and Giangiacomo Minak

Department Industrial Engineering, University of Bologna, via Fontanelle 40, 47121 Forlì, Italy; ana.pavlovic@unibo.it (A.P.); davide.sintoni@studio.unibo.it (D.S.); giangiacomo.minak@unibo.it (G.M.)

* Correspondence: cristiano.fragassa@unibo.it; Tel.: +39-347-6974046

Received: 16 March 2020; Accepted: 10 April 2020; Published: 12 April 2020

Featured Application: A photovoltaic roof for vehicles, in sandwich-structured composite, is designed for optimizing static stiffness and dynamic response, but also energy efficiency.

Abstract: A multi-step and -objective design approach was used to optimize the photovoltaic roof in a multi-occupant racing vehicle. It permitted to select the best combination of design features (as shapes, widths, angles) in composite structures simultaneously balancing opposite requirements as static strength and dynamic stiffness. An attention to functional requirements, as weight, solar cells cooling and solar energy conversion, was also essential. Alternative carbon fiber-reinforced plastic structures were investigated by finite elements using static and modal analyses in the way to compare several design configurations in terms of natural frequencies, deformations, flexural stiffness, torsional stiffness, and heat exchange surfaces. A representative roof section was manufactured and tested for model validation. A significant improvement respect to the pre-existing solar roof was detected. The final configuration was manufactured and installed on the vehicle.

Keywords: design optimization; solar vehicles; photovoltaic roof; lightweight structures; carbon fiber-reinforced plastic (CFRP); natural frequencies; stiffness; heat exchange; Ansys ACP

1. Introduction

The design process in engineering is often based on a cyclical path aiming to improve existing solutions. However, this step-by-step approach becomes rather complex to be applied when the need exists to consider different optimization criteria at the same time. Such a situation can occur in many real applications [1], but it comes to be essential in the case of competition vehicles.

Sports car design surely represents one of the most exciting test benches for designers to set up efficient design methodologies and solutions. The desire of competitiveness commonly drives the designer to direct the design action towards a multi-objective optimization (MOO).

In the case of a Formula One car [2], for instance, the opportunity to lighten the vehicle comes up against the need to offer higher lift [3] and, then, requires a proper balance between weights and loads [4]. In these terms, the field of solar races is maybe even more interesting, where the power involved is extremely low and the optimization criteria must be even more shrewd [5].

Solar cars are electric vehicles where the energy for powertrain is provided by the Sun thanks to the installation of photovoltaic panels (usually on the roof) and a battery pack for energy storage. Thus, energy efficiency definitely represents the key concept driving each design choice in the case of solar prototypes [6]: aerodynamics [7], dynamics [8], kinematics [9] and even manufacturing [10].

However, weight minimization probably stands for one of the most deemed design policies with the scope to minimize the inertial masses and energy losses. In accordance with [11] based on electric cars, in fact, an amount of 13% can be saved in term of energy if a 10% decrease in weight is achieved. Weight reduction strategy in vehicle design is often associated to the massive use of composite materials

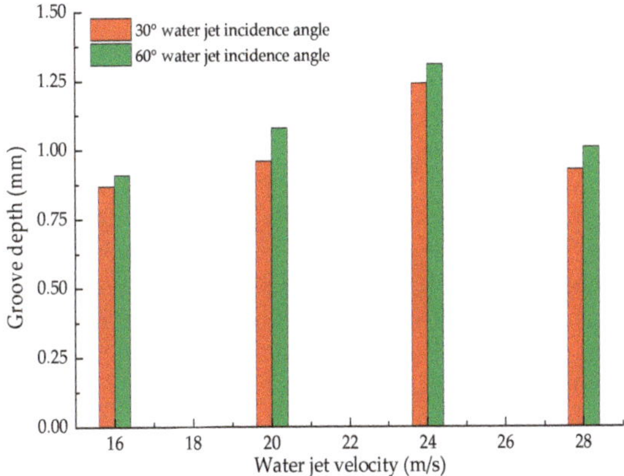

Figure 6. Groove depths obtained at different water-jet velocities (horizontal axis) and impact angle (colors).

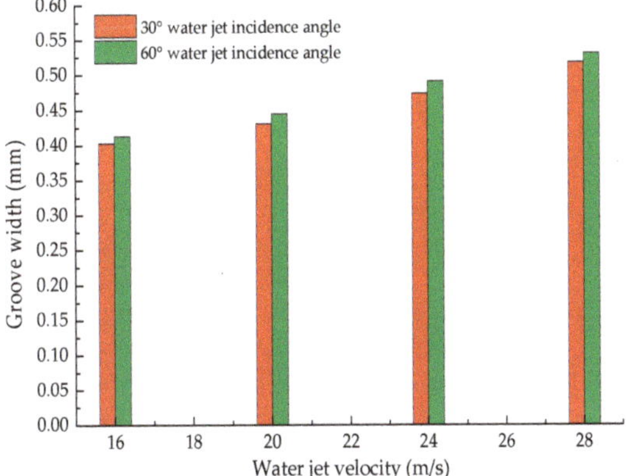

Figure 7. Groove widths obtained at different water-jet velocities (horizontal axis) and impact angles (colors).

As shown in Figures 4–7, the depth and width of the laser-etched grooves in the polysilicon material are dependent on water-jet velocity but not impact angle (the water-jet velocity trends were very similar at both 30° and 60°). The depth of the processing groove initially increased with increased jet velocity from 16 to 24 m/s but decreased between 24 and 28 m/s. In addition, the width of the processing groove gradually increased across the entire range of tested jet velocities. For the same water-jet velocity, the groove depths and widths were slightly larger at an impact angle of 60° than at 30°. The slight increase at 60° is attributable to the higher impact force at 60° than at 30°. As the water-jet velocity increased from 16 to 24 m/s, both the impact and cooling effect of the water-jet increased; however, the increasing trend of the cooling effect slowed down. As a result, the total increasing trend was dominated by the impact of the water jet, whereas the cooling effect played a secondary role. The molten silicon and slag deposited on the bottom of the tank were removed to reduce laser energy absorption; thus, the depth and width of the processing grooves increased gradually with water-jet velocity. However, at a water-jet velocity of 28 m/s, the large impact force was compromised because the experimental device was disturbed. At this velocity, the impact of the water jet deviated from the center of the jet impact, and the water beam diverged when striking the material surface, which caused splashes and "water mist" that affected the focused laser spot. In water-assisted laser processing, the material surface is covered with a thin, flowing layer of water, which absorbs some of the laser energy, thereby weakening the laser's action on the material. Simultaneously, the like carbon fiber-reinforced plastics (CFRP) that allow us to make lightweight structures [12]. Thanks to CFRP, the designers have at their disposal two specific tactics for structural optimization [13,14], directly linked to the mechanic properties of composites and able to permit lighter structures: (1) high stiffness to weight ratio; and (2) strong anisotropy in the mechanical properties.

The possibility to design the material properties in different directions leads to a greater designers' opportunity in optimization. However, as mentioned, this optimization may involve multiple aspects, not always related to the same category. Precisely for this reason, the general concept of multi-objective optimization is spreading in vehicle engineering [15].

The scientific term of 'optimization' refers to that branch of applied mathematics which studies theory and methods for the research of the maximum and minimum points of a mathematical function [16]. The scope is to translate into mathematical terms a given problem which can be related to different disciplines, as physics, finance, social sciences and, of course, engineering [17–28].

In [18] and [19], for instance, the structural improvements in vehicle components using a design procedure not more based on a trial-and-error process, but linked to optimization methods are discussed, showing their potential, in the case of weight reduction in automotive chassis design.

Far from intending to represent all the general aspects of the optimization problem or its methods and processes in engineering (in part available in [24]), it is possible to here summarize some fundamental elements of specific interest for the present study, as:

- The optimization problem aims to find the optimal solution from all feasible solutions,
- Where variables can be continuous or discrete, creating a divergent solution space;
- It assumes that, thanks to classification mechanisms, it is possible to find the optimal solution without checking all the possible combinations, one by one;
- The optimization criteria, at the same time, must be expressible in terms of functions of different variables, including relative minimum and maximum limits;
- It is usually advisable to standardize and make comparable these outputs by factoring them (e.g., by max value) in the way to provide unitary indexes (scores between 0 and 1);
- In the simplest case the problem is scaled down in maximizing (or minimizing) a real function by choosing the input values from a set allowed and calculating that function.

Although widely used in engineering (as for structural design, product design, shape optimization, topology optimization, processing planning, and so on), the complexity of applying optimization methods as generally developed by mathematicians has often convinced vehicle designers to additionally consider hybrid and/or simplified approaches.

In [25], for instance, optimized solutions in structural design were achieved merging the Taguchi's method for robust design and the particle swarm algorithm for optimization.

The general state of the art in the field of composite laminates and sandwich optimization is available in two recent comprehensive reviews [26,27]. Results from analyses carried out since 2000 are discussed based on the type of structures, objective functions, design variables, constraints and applied algorithms. Addition factors as boundary conditions, orientation of fibers, design variables are also considered respect to improvements in mechanical behavior such as buckling resistance, stiffness and strength along with reducing weight, cost and stress under various types of loadings.

Closer to the current case, in [28] an aerodynamic shape optimization with the scope to reduce the drag coefficient is performed based on a hybrid process that coupled a genetic algorithm with a quite common iterative method for solving unconstrained nonlinear optimization problems. As often happens, this investigation merged potentialities of a commercial code (a Computational Fluid Dynamics code in the case) with optimization methods and validation test cases.

Furthermore, in [29] a simpler methodological approach is proposed and used to develop an automotive structure able to balance technological, economic and ecological aspects at the same time. It is focused on the application of foams as a core material in sandwiches for floor panels in a concept car. The problem of multi-objective optimization is solved by a procedure of weight minimization that also ponders the structure behavior respect to static loads in terms of stiffness, strength and buckling constraints. The comparison is formulated for each material application, including considerations on mass and materials' price, and environmental impact.

Strictly in line with that last two cases, the current work was based on a multi-objective optimization where a compound objective function was adopted in the form of a sum of performance indexes balanced by proper weighting coefficients. The specific scope of the paper and its novelty are to show the development of a practical real-life application from the stage of the conceptual design till the actual embodiment and the manufacture of the component. In this framework, the multi-objective optimization is constrained from the beginning by geometrical and functional requirements, moreover also the raw material and the manufacturing technology cannot be changed.

2. Materials and Methods

2.1. Composite Materials and Structures

This study aimed to redesign a new photovoltaic roof for the solar vehicle, designed and built by the University of Bologna to take part in solar competitions worldwide [30].

The vehicle is a four-passenger quadricycle [31] where every technical solution was designed with the scope to improve the overall car performance (Figure 1). The replacement of the out-of-date photovoltaic panels, as well as changes in the safety structure eliminating the metal roll bar [32], driven by the need to respect rules of a different race, gave the chance to intervene on the roof through this optimization.

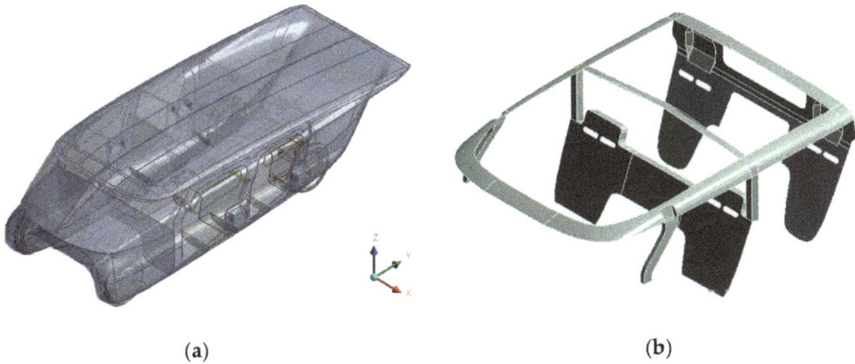

(a) (b)

Figure 1. The multi-occupant solar vehicle: (a) general layout; (b) roof support structure.

As a racing car, each part of the vehicle should be as light weight as possible, but also with excellent mechanical properties, permitted by the large employment of high stiffness and strength Carbon fiber reinforced plastics (CFRP). The same occurred for the roof, which is the structure of the photovoltaic panel, where unidirectional (UD) and bidirectional (twill) CFRP were layered in the form of sandwich structured laminates. Specifically, Toray T1000 UD and T800 twill pre-impregnated fabrics with, respectively, 0.15 mm and 0.30 mm thicknesses, were chosen considering their remarkable mechanical properties. Polyvinyl chloride (PVC) foam is used as sandwich core. The outermost layer is a T800 fabric, with fibers along 0°/90° direction. Internally there are two unidirectional layers in T1000, at 0°. The PVC core is not present everywhere, but only in the central section, having a roof with variable thickness, between 1.0 and 10.0 mm (and 4.0 mm in this case). Table 1 reports the main mechanical properties of materials and Table 2 their layout. It is evident the marked anisotropy (unidirectional or bidirectional) in properties of composites, but also a symmetrical layout respect to the core.

With reference to the direction of the reinforcing fibers, it should be noted that the angle in the table is to be considered with respect to the construction method (better specified below). In grid structures the main directions (0°) are along the direction of the different beams. In the perforated panels there is a single main direction that coincides with the longitudinal direction of the roof.

Table 1. Main material characteristics and properties.

Property	Unit	T1000	T800	PVC
Type	-	Unidirectional	Twill	Foam
Density	Kg/m^3	1490	1420	100
Young's Modulus	MPa	121,000, 8600, 8600	61,340, 61,340, 6900	125
Poisson's Ratio	-	0.27, 0.4, 0.27	0.04, 0.3, 0.3	0.40
Tensile Stress Limit	Mpa	2231, 29, 29	805, 805, 50	2.5

Table 2. Composite layout.

Layer	Material	Thickness	Angle
1	T800	0.30 mm	0°/90°
2	T1000	0.15 mm	0°
3	T1000	0.15 mm	0°
4	PVC	4.0 mm	–
5	T1000	0.15 mm	0°
6	T1000	0.15 mm	0°
7	T800	0.30 mm	0°/90°

2.2. Optimization Parameters

In general terms, the design optimization of the roof in the case of this solar vehicle was based on a combination of mechanical and functional features, as reported in Table 3. The first ones, i.e., the mechanical features, are related to the structural stiffness respect to static and dynamic loads and are required to assure the best conditions in terms of integrity, safety and modal response. They can be measured by flexural stiffness, torsional stiffness and (first) resonance frequency in the way that the higher their values, the better for the design solution is. The second ones, i.e., the functional features, are also desired to assure the best performances in terms of energy efficiency, including the reduction of inertial masses and possibility of maximizing the area exposed to heat exchange.

Table 3. Optimization objectives and related weighting coefficients.

Features	Output	Unit	Target	Weighting
Mechanical	Flexural stiffness	N/mm	Highest	0.15
	Torsional stiffness	N·mm/rad	Highest	0.15
	Resonance frequency	Hz	Highest	0.30
Functional	Heat transfer surface	mm^2	Highest	0.40
	Weight	Kg	Lowest	(fixed)

At the same time, since these features are clearly interconnected, it was preferred to reduce the size of the system by fixing one parameter, the weight (= 0.250 ± 0.010 kg) as common design target, and performing the optimization based on the other factors. Since each factor has to be maximized, it was possible to express the performance index, the index for comparison (IC) as:

$$\mathrm{IC} = \frac{1}{\sum_{i=1}^{n} p_i} \sum_{i=1}^{n} p_i \left(\frac{y_i - y_{i\,max}}{y_{i\,min} - y_{i\,max}} \right) \quad (1)$$

where p_i represents the weighting coefficients (with $i = 1 \ldots 4$ and $\Sigma\, p_i = 1$) and y_i is the i-th optimization parameter which can vary between y_{imin} and y_{imax}, representing the minimum and maximum value assumed by each parameter y_i.

2.3. Overall Methodology

The present investigation was also based on the following concepts and phases (Appendix A):

1. Preliminary shape definition when different geometrical shapes were considered, specifically rectangles, ellipsis and triangles, in terms of their ability to minimize the surface.
2. Manufacturing techniques taken into consideration since the very beginning, identifying two practical solutions in production and four geometrical options (Figure 2):
   - A composite laminate made in a single piece where rectangular holes, with rounded edges (a), or elliptical ones (b) where shaped through;
   - A grid-based structure, made up of a series of intersecting straight (vertical, horizontal and angular) lines (grid lines), forming a rectangular (c) or triangular (d) texture of beams.
3. Geometric shape (topology) optimization performed by automatic algorithms using screening and response surface method respect to the most relevant geometric parameters (e.g., axes for ellipsis, lengths for rectangle, distances between grids (as reported in Table 4)).
4. Score criteria, as flexural and torsional stiffness or first resonance frequency, derived by finite element analyses (FEA) using commercial codes in static and modal simulations.
5. Valuations performed, at first, respect to a basic unit, dimensionally set to a 500 × 500 mm section (roughly equivalent to 4 × 4 solar cells).
6. An overall structure for the roof made up by a repetition of this basic unit.
7. The basic unit solution, as here optimized, adopted to produce a larger section, a 2500 × 800 mm (~2 m^2) flat mock-up, to be used for experimental (modal) test and numerical model validation.
8. This design solution also applied to the roof shape in accordance with its real double curvature and the 3D model (~5200 × 1.600 mm) then analyzed by FEA respect to static and dynamic loads.
9. The roof definite structure manufactured by autoclave composite techniques, installed on the solar car and finally examined in real operative conditions.

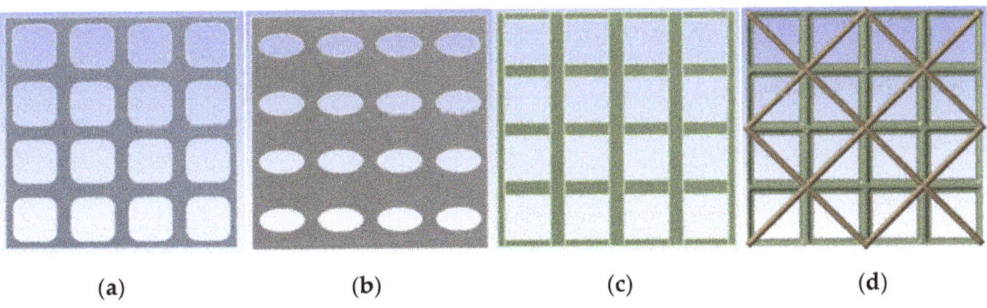

(a) (b) (c) (d)

Figure 2. Shape definition: four geometries were considered: (**a**) rounded rectangles and (**b**) ellipses holes, or (**c**) rectangular and (**d**) triangles in quadridirectional grids.

It is noteworthy that, actually, Table 4 reports only a subset of parameters suitable to describe in their generality the geometric properties of the represented shapes (i.e., rounded rectangles, ellipses, rectangular and triangles) especially as regards the grid configurations since the possibility to overlap beams in a large variety of combinations (angles, distance between elements' recurrences, etc.).

Table 4. Geometrical parameters for optimization.

Rectangular Hole	Elliptical Hole	Rectangular Grid	Quadridirectional Grid
Width (***W***) Height (***H***) Fillet radius (r)	1st axis (*A*) 2nd axis (*B*)	1st Width (***a***) 2nd Width (***b***)	1st Width (***a***) 2nd Width (***b***) Angle (***α***)

2.4. Optimization Space Reducing

However, these options already create a quite large *n*-dimensional optimization space (with n ≥ 10 based on how certain geometric similarities are taken into account) which must be deemed by four optimization objectives (output), or even five, if the weight is reintroduced in the comparison. Each point of this *n*-dimensional space should represent a potential solution of the optimization problem, that deserves to be investigated (in terms of stiffness, resonances and so on) by three separate numerical analyses (one modal and two static). At the same time, it is evident it is not possible to reiterate these calculations for an amplified number of situations. Then, as commonly happens in each optimization analysis, a way to decrease the size of the solution space was achieved by:

- Decoupling the effects related to the geometric characteristics of the basic shapes (i.e., rectangle, triangle, etc.) from those related to their recurrence and rearrangement. It was obtained repeating the optimization procedure acting on different levels. Specifically, in Figure 3 these different scales of analysis are displayed, showing the basic unit (500 × 500 m), used for geometrical optimization of shapes (Figure 2), and the mock-up (2500 × 800 mm), used for grid optimization, experiments and model validation. It should be noted that the figure shows half (front) section of the vehicle roof (with ~5200 × 1.600 mm as overall dimension);
- Limiting the (full) parametric (and automated) analysis to an optimization only based on the structural outputs as first; the most promising solutions from this topologic optimization are then (individually) verified in terms of impacts on the performance outputs.

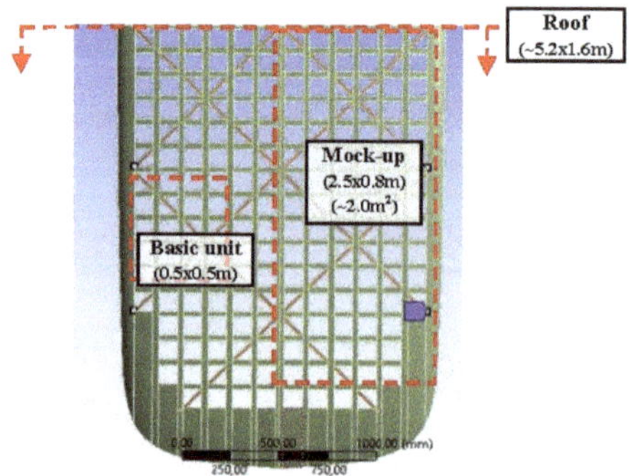

Figure 3. Different scales of analysis: basic unit (500 × 500 m), experimental mock-up (2500 × 800 mm) and half (front) section of the roof (~5200 × 1.600 mm).

2.5. Design and Simulation Tools

3D CAD modelling was performed in SolidWorks (by Dassault Systèmes, Vélizy-Villacoublay, France). Ansys Workbench ver. 18.1 platform (by ANSYS, Inc., Canonsburg, Pennsylvania, USA) was used for structural analysis of composites (by an ACP toolkit) and for design optimization (Design Exploration) by the direct optimization features and the screening and response surface method.

Specifically, in 3D modelling solid parts were suppressed with the scope to manage 3D surfaces. Discretization was done by shell elements (FE), quadrangular of quadratic order with 8 knots, preferred for precision. After mesh convergence tests, the maximum size for FE was set at 3 mm (in the case of basic units), employing 12,000–18,000 FE for meshing the different configurations.

In accordance with reality, the thickness of sandwiches was made variable along the section by creating cut-off selection rules in ACP and assigning them to the PVC core. The same technique was adopted to manage edges and intersections. A visual example of effects of expedients used for the correct discretisation of composite structures is reported in Figure 4 in the case of rectangular grid. The 500 × 500 mm base section, consisting of 4 × 4 squares (representing the solar cell frames) are shown highlighting differences in thickness and the final mesh with implemented these changes.

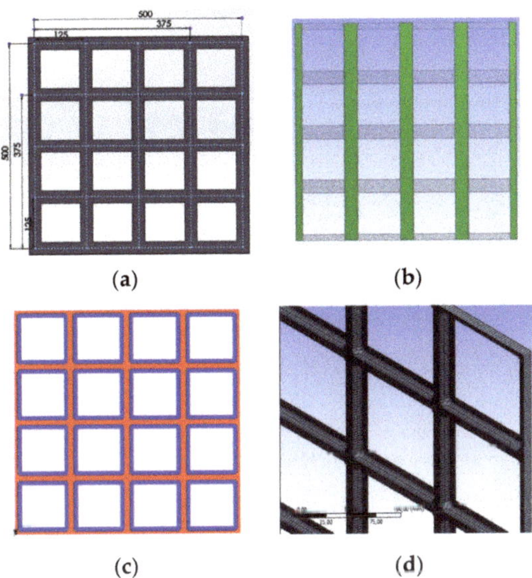

Figure 4. Details in sandwich discretization (in the case of rectangular grid): (**a**) the 500 × 500 mm base section; (**b**) vertical beams over respect to the horizontal ones; (**c**) sides where the thickness of the sandwich must decrease; (**d**) final mesh (also showing the thickness changes).

The so-called single-layered shell method [33], instead of many others (e.g., stacked shell [34]), was chosen and implemented for discretizing the layout. This simplification quite common in analyzing composites in the case of large and complex structures, permits to quickly investigate the main phenomena at the level of macroscale. With a proper conversion in properties, it reduces a multilayered laminate to an equivalent single layer laminate and use shell elements all along the surface with integration points (IP) throughout the thickness [33]. Specifically, in the case an IP was set per each of the seven layers: no additional IP was used to investigate the six interfaces between layers. In such a way it was possible to drastically limit the number of FE speeding up the simulation focusing the attention on in-plane phenomena [33–35].

However, this methodological limit has no practical effect on the present study since the roof has no global structural functions [35]. The vehicle was designed around a lower monocoque and an upper structure (Figure 1b). A rigid frame on the composite structure guarantees the fixing and solidity of the roof which, practically, in addition to its own weight, must only support the solar cells (~ 1.2–1.5 kg/m^2).

From the Ansys internal library, material models labelled as "Epoxy Carbon UD (230 GPa) Prepreg", "Epoxy Carbon Woven (230 GPa) Prepreg" and "PVC Foam (80 kg m^{-3})" were chosen for, T1000, T800 and PVC respectively, but default properties were changed in accordance to Table 1. ACP toolkit permitted to build exact layouts as shown with Table 2. In the grid configuration, in order to correctly orient the fibers according to the direction of the single beam, it was necessary to divide the grid in section subgrouping beams characterized by a same orientation in the fibers.

2.6. Simulation Procedures

Three different simulations were carried out on each configuration under investigation.

2.6.1. Resonance Frequency

The first simulation was a modal analysis and concerned the resonance frequency. It was carried out in free-free conditions: no constraints or forcing were set. The first six modes are rigid modes, corresponding to the structure's six degrees of freedom. Therefore, the seventh mode and frequency correspond to the first modal shape of interest (Figure 5a).

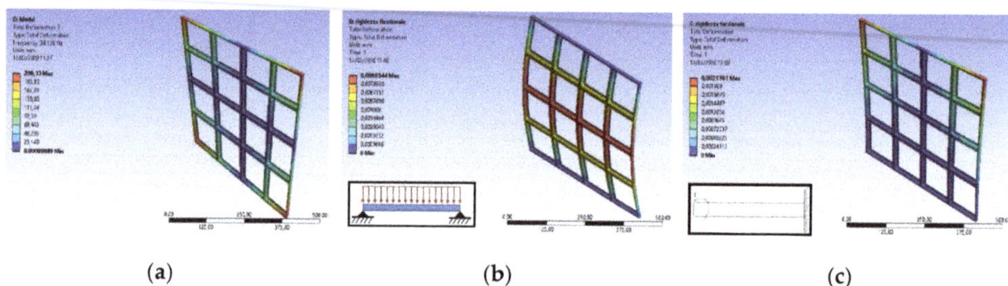

Figure 5. Numerical evaluation of (**a**) resonance frequency, (**b**) flexural and (**c**) torsional stiffness.

2.6.2. Flexural Stiffness

The second simulation was a static analysis and concerned the flexural stiffness. Load conditions were set equivalent to a simply supported beam with distributed load. Constraints were assigned to two opposite external sides of the beam and a force equal to 1 N (to simplify calculations) to upper face. The model was simplified as a beam considering the complete roof structure, consisting of various repetitions of it, has a prevalent dimension over the others. The required output is the flexural displacement (total deformation). Flexural stiffness can be calculated by dividing the force by the maximum displacement (Figure 5b).

2.6.3. Torsional Stiffness

The third simulation was a static analysis, as the previous one, but concerned torsional stiffness. As a system, it was considered a cantilever beam with a pure twisting moment applied at the end. A fixed support was therefore assigned to one side and a torque of 1 N·mm to the opposite side. Torsional stiffness can be calculated as:

$$K = \frac{T}{\theta}$$
$$\theta = tan^{-1}\left(\frac{z_1 + z_2}{2b}\right)$$

where T is the twisting moment, θ is the rotation, z_1 and z_2 are the vertical displacements at the two ends and b is the length of the beam. The rotation θ was determined, in the case, directly using the

Flexible Rotation Probe function instead of the analytical formula after verified the results by the two methods were similar (Figure 5c).

3. Results and Discussion

3.1. Shape Topological Optimisation

In Table 5 a first comparison between geometries at the level of basic unit is reported in terms of performance index (IC) respect to their main dimensional characteristics (sizes).

Table 5. Effect of changes in geometry (at the level of basic element).

Configuration		Laminate		Grid			
		Rectangle	Ellipsis	Orthogrid	Cross	Double cross	Quadridirection
Hole Size (mm)		94.7 92.8 23.7	95.0 95.0				
Grid Width Size	Upper grid (mm) Lower grid (mm)			40.7	30.0	30.0 23.5	30.0 14
Lowest Frequency (Hz)		84.2	73.6	64.3	88.9	97.3	111.9
Flexural stiffness (N/mm)		209	103	204	137	114	122
Torsional stiffness (N mm/rad)		5247	3193	4391	5170	5319	6623
Element Weight (kg)		0.263	0.263	0.264	0.263	0.265	0.264
Performance Index		**0.61**	**0.10**	**0.32**	**0.48**	**0.53**	**0.79**

Specifically, in the case of laminates dimensions refers to the hole sizes as width (W), height (H) and fillet radius (r) for a rounded rectangle or 1st and 2nd (A, B) axes for ellipsis.

In the case of the grids, the dimensions refer to the width of beams that make up the structure: an upper orthogonal (0–90°) grid, reinforced by a lower diagonal (⊥45°) one. A value of 30mm in width for the upper grid is fixed considering the need to sustain the singular solar cell. Both for laminates and grids 5.2 mm of thickness is fixed since the predetermined composite layout (Table 2).

In terms of results, it should be noted that, together with the torsional and flexural stiffness, the first (lowest) natural frequency is also reported in the table. This value represents an indication of how sensitive the structure is to the effects of vibrations. The maximum displacement respect to the first natural frequency should be also considered representing the effect of vibrations in terms of intensity (not sensibility).

By comparing the values of the performance index (IC), it is immediately evident that the best configurations are, in general, grid configurations (as Figure 2c,d) reinforced by several crosses (IC ≥ 0.48) and, specifically, a quadridirectional configuration (IC = 0.79). Despite this, the option of manufacturing a laminate in a single body and then lightening it through holes could be taken into consideration provided the choice of rectangular (IC = 0.61), almost squared (~95 mm vs. 93 mm), and rather large holes (~8800 mm^2) with rounded borders (~24mm). In making these considerations, it should be also remembered that all these configurations are obtained keeping the same weight (w = 0.264 ± 0.01 kg) as target and, consequently, the same amount of material.

Outcomes from Table 5 also take into count, as a preliminary optimization (improved below), the way an upper orthogonal grid (orthogrid) can be reinforced by lower transversal crosses leading to double grid configurations. The most interesting options, between many others under consideration, are displayed in Table 6 where at the investigation level of the basic unit, the impact of a crescent number of crosses is analyzed. The same table also exhibits the effect of changes in cross angles.

Table 6. Effect of changes in the crosses (at the level of basic unit).

Configuration		**Number of Crosses**								
		1	$1^1/_3$	$1^1/_3$	$1^1/_2$	$1^1/_2$	2	2	4	4
Width Size	Upper grid (mm)	30	30	30	30	30	30	30	30	30
	Lower grid (mm)	30	28	28	23.5	23.5	20	20	10.5	10.5
Load Direction		x or y	x	y	x	y	x	y	x	y
Lowest Frequency (Hz)		88.9	83.5	83.6	97.3	89.8	84.7	84.9	62.7	63.9
Flexural stiffness (N/mm)		137.1	127.9	121.4	114.0	116.2	169.8	113.9	194.4	115.4
Torsional stiffness (N mm/rad)		5170	4676	4248	5319	4604	4786	4459	3675	3219
Element Weight (kg)		0.263	0.266	0.266	0.265	0.265	0.266	0.266	0.266	0.266
Performance Index		**0.68**	**0.52**	**0.45**	**0.75**	**0.56**	**0.68**	**0.47**	**0.30**	**0.02**

Results, as said, assumed that:

- All beams in each specific grid have the same dimensions (i.e., width and thickness);
- These dimensions can vary (in general) between the upper and lower grids;
- The thickness, however, is fixed by the specific composite layout;
- The width of beams in the upper grid is fixed (with the scope to permit to sustain the solar cells);
- The width of beams in the lower grid is related to their number (since a constant weight).

For each configuration, the table reports stiffness in the cases of application of (flexural or torsional) loads along X or Y axes. Differences are evident related the geometrical anisotropies of grids respect to these changes in the axes.

3.2. Grid Topological Optimisation

In Table 7, a second comparison between configurations is reported. It also deals with the best solutions detected in terms of geometry and shapes at the level of basic units (Table 5) but extends results on the mock-up case. In particular, the laminate with rectangular holes where compared with quadridirectional grids in different configurations. Having demonstrated a certain appropriateness (measured through IC) for a specific configuration at the level of basic unit does not necessarily mean, in fact, that this rank remains the same on the mock-up. Such a fact immediately emerges from the table making clear that the laminate with rectangular holes (IC = 0.25) is no longer convenient.

Having also demonstrated the convenience of using, as general design concept, a grid structure where an upper orthogonal grid (on which the solar panels are laid) are reinforced by a series of diagonal crosses, a different topological optimization study was carried out with respect to the definition of the (lower) grid in terms of number of elements and their arrangement.

Without entering in unnecessary details, Table 8 exhibits the impact of diagonal crosses at the level of mock-up, both as number, distribution and crosses angles.

Table 7. Effect of changes in grid shape (at the level of mock-up).

Configuration	**Laminate**	**Quadridirectional Grid**		
	Rectangle	**Low**	**Medium**	**High**
Lowest Frequency (Hz)	11.4	12.3	12.9	13.8
Flexural stiffness (N/mm)	7.7	4.4	3.5	3.6
Torsional stiffness (N mm/rad)	1851	1960	2090	2665
Element Weight (kg)	2.090	2.047	2.031	2.028
Performance Index	**0.25**	**0.27**	**0.38**	**0.76**

Table 8. Effect of changes in grid topology (at the level of mock-up).

Configuration		**Quadridirectional Grid**						
Crosses		**2 × 1**	**$2^{1}/_{3} \times 1^{1}/_{3}$**	**$2^{2}/_{3} \times 1^{1}/_{3}$**	**$3^{1}/_{2} \times 1^{3}/_{4}$**	**4 × 2**	**$5^{1}/_{4} \times 2^{1}/_{4}$**	**8 × 4**
Width Size	Spacing (mm)	1000 × 1000	875 × 875	750 × 750	625 × 625	500 × 500	375 × 375	250 × 250
	Upper grid (mm)	30	30	30	30	30	30	30
	Lower grid (mm)	50	42	38	33	30	22	14
Lowest Frequency (Hz)		14.5	13.5	14.7	14.0	12.9	15.9	13.8
Flexural stiffness (N/mm)		3.4	4.5	4.3	4.3	3.5	3.7	3.6
Torsional stiffness (N mm/rad)		2587	2252	2379	2358	2090	2156	2665
Element Weight (kg)		2.031	2.031	2.031	2.031	2.031	2.033	2.028
Performance Index		**0.49**	**0.42**	**0.64**	**0.50**	**0.03**	**0.60**	**0.46**

The beam width in upper grids remains 30 mm (to conveniently host the solar cells), but changes affect the lower ones (from 14 to 50 mm) according to the condition of weight conservation. Spacing parameter (in Tables 8 and A2) provides information on the distance between beams on the lower grid: reducing the distance increases the number of beams (and crosses as reinforce) but, on the other side, as said, reduce their width saving the total weight. Finally, unlike the first analyses on the geometric parameters, where an automatic change for (almost) all the parameters was possible, in these further investigation configurations often involve the development of specific CAD models. Thus, several actions were done manually.

3.3. Multi-Objective Optimization Results at a Glace

In line with results from similar investigations (as [36]), it is clear how the grid structures are preferable to perforated panels. It is also evident that a grid with a denser mesh is better than those of sparse one with an equivalent weight. However, although the comparisons give unambiguous indications when only mechanical properties are considered, additional criteria should be included.

In fact, the optimized option as here detected, i.e., quadridirectional grid with crosses every 250 mm, could present (compared to less dense grids) additional challenges in its applicability such as:

- Minor free surface for heat transfer, not permitting an efficient the solar cells cooling;
- Greater complexity in fabrication related to a larger number of basic elements (beams).

These aspects were not including in the optimization since the beginning and are not considered.

A preliminary estimation of the free surface carried out by Ansys with respect to two borderline grids showed a quite low variability (<10%): using a denser grid does not significantly affect the panels cooling. It also depends to the fact that, denser grids are necessarily made by tighter beams (since the precondition of equivalent weight).

Regarding the composite roof construction, with an area of approx. 8.0 m^2, its manufacturing can represent a laborious task. A larger number of beams and crosses, to be made and glued, would be preferable only when mechanical properties were significantly better: it is not the present case.

Table 9 reports an update in design solutions evaluation, taking into count of additional parameters and objectives of optimization. In particular, the maximum displacement when the lowest natural frequency occurs is included: those parameters have to consider in combination for better analyze the dynamic behavior of the structure. Furthermore, the table also introduces parameters, not strictly related to mechanical properties, as the area available for heat transfer and an indicator of producibility. While the exposed area may be directly detected by Ansys functionalities, an empirical index had to be defined for estimating the producibility, it was done in accordance with information from manufacturer, making its value proportional to the number of operations necessary to build composite structures.

Table 9. Multi-objective optimization of composite structure.

Configuration				**Laminate**	**Quadridirectional Grid**		
	Weight	**Target**	**Best**	**Rectangle**	**Low**	**Medium**	**High**
Flexural stiffness (N/mm)	0.25	←	7.7	7.7	4.4	3.5	3.6
Torsional stiffness (N mm/rad)	0.10	←	2665	1851	1960	2090	2665
Max displacement (1st frequency) (mm)	0.10	↓	61.6	61.6	63.0	62.6	70.4
Lowest Frequency (Hz)	0.10	←	13.8	11.4	12.3	12.9	13.8
Heat transfer area (m^2)	0.15	←	1.1	1.07	0.83	0.83	0.85
Producibility	0.30	↓	1.0	0.75	0.80	1.00	0.50
Element Weight (kg)	—	↓	2.090	2.090	2.047	2.031	2.028
Performance Index	**1.00**		**0.57**	**0.50**	**0.39**	**0.57**	**0.37**

As a synthesis, a quadridirectional grid with crosses every 500 mm was adopted. In accordance with previous results, in fact, this structure exhibited good mechanical properties, better than both perforated sandwich panels (especially regarding the resonance frequency [37]), but also compared to an orthogonal grid, thanks to the presence of reinforcing beams in the diagonal directions [38]. However, its medium sparse grid does not entail the construction problems that can occur during production in the case of denser grids.

Finally, since it was noticed that a slight increase in the mass of the angular grid provided positive benefits to the quadridirectional grids, it was also preliminary checked the effect of minor changes in the constraint of the minimum width for the orthogonal beams (=30 mm).

The basic unit of 500 mm sides was examined, performing a further optimization in Ansys by varying the width of the orthogonal beams between 25 mm and 30 mm and diagonal beams between 30 mm and 50 mm. Since the condition of equal total weight, an increase in the mass of one grid, orthogonal or diagonal, decreases the other. In general terms, the best situation is present when widths of both grids are quite similar. A specific optimization can be obtained when the choice of the solar cells, defined the lower value of width for elements in the upper grid.

3.4. Validation

A modal analysis was used for validating the FE model by experiments.

The mesh consisted of 56163 shell elements and allowed to identify the first 20 modes of the structure (of which, six are rigid modes). Deformations along main directions and Cartesian axes were also evaluated. On the other side, a modal test was also carried out both with accelerometers (as in [39–41]) and Bragg grating fibers (FGB) sensor (as done in [42]). Figure 6 and Table 10 summarize this comparison for a range lower than 100 Hz, equivalent to the five lowest natural frequencies. A good accuracy is clear with a 7.3% average deviation between predictions and experiments. Besides, a constant underestimation is also evident which suggests a refining in the FE model discretization.

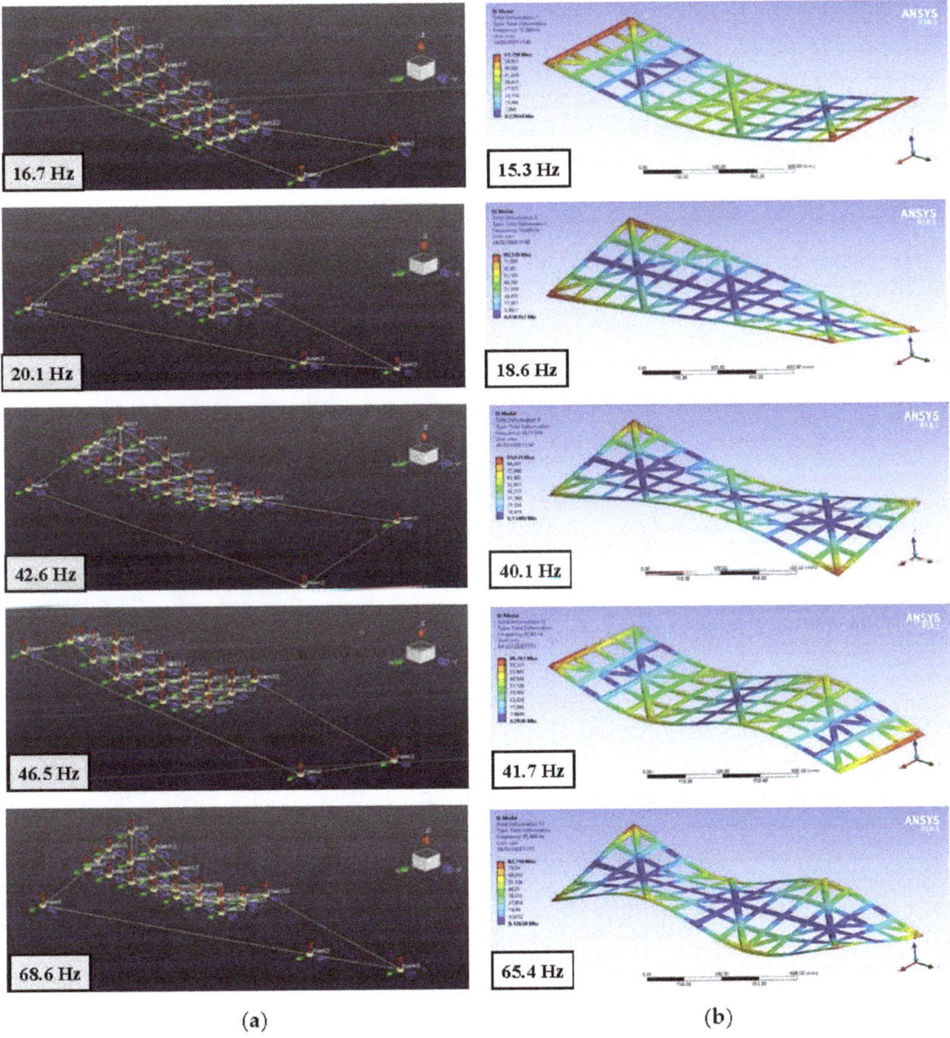

Figure 6. Modal analysis with results from (**a**) experiment and (**b**) simulation.

Table 10. Comparing the results from modal analysis in terms of natural frequencies.

Mode	Frequency (Hz)		Difference (%)
	Measure	Simulation	
I	16.7	15.26	−8.6%
II	20.1	18.69	−7.0%
III	42.6	40.13	−5.8%
IV	46.5	41.67	−10.4%
V	68.6	65.40	−4.7%

3.5. Results Implementation

The same design approach was used to finalize the roof design. The 3D geometry of the roof was filled by structural modules with design parameters based on the previous optimization. Specifically, a quadridirectional grid with a medium density of beam element was used as valid compromise between the different targets. It was characterized by an orthogonal grid on the top and a lower diagonal grid on the bottom with, respectively, 300 mm and 200 mm widths. This design pattern was replicated along the roof dimension and shape, partitioned in front and back sections (Table 11).

Table 11. Overall geometrical dimensions for roof sections.

Section	Dimensions [m]	Surface [m^2]
Front	1.714 × 2.344 × 0.142	4.567
Back	1.540 × 2.056 × 0.510	3.522
Total	1.714 × 2.344 × 0.510	8.089

The dynamic behavior of these parts was examined by a further modal analysis considering unloaded bodies (as previously) but constrains as in real case. First frequencies were 21 and 12 Hz with total displacements of 42 and 69 mm for, respectively, front and back sections (Figure 7)

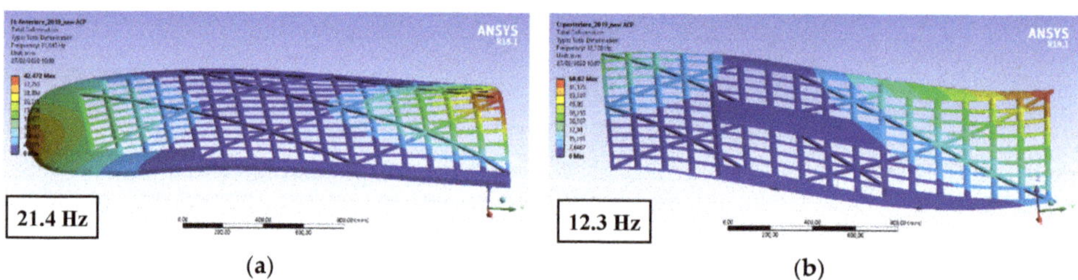

(**a**) (**b**)

Figure 7. Numerical modal analysis of (**a**) front, (**b**) back section of the roof.

Then, the composite structure was manufactured using manual layout and autoclave molding and hot gluing with two-component for part assembly. The solar cells, E60 bin Me1 by Sunpower, were directly laminated on the panel with ethylene vinyl acetate (EVA) films. Flexible layers on both surfaces (front and back) for a total of five layers (including 2 EVAs), for a 1.5 mm of overall thickness, were overlapped and cured in autoclave. In particular, solar cells were positioned with the scope to maximize the energy yield of vehicle-integrated photovoltaics (VIPV) [43].

Lastly, the solar roof was installed and finalized with other vehicle components. Figure 8 shows several imagines from the solar vehicle at the end of the present investigation. Functional tests were performed running over 800km on roads open to traffic. No complications emerged related during this first trial in terms of structural design of the roof which appeared stable and functional.

Figure 8. Solar roof manufacturing and installation on the solar vehicle.

3.6. Further Considerations and Novelty

Following the use of a multi-objective optimization in this design action, several general considerations can be introduced. As first, it is essential to note how the outcomes strongly depends on the weight assigned to the objectives: even minor changes in their values can lead to very different design indications. Thus, it would be relevant to find criteria for an empirical definition of these parameters. By an analysis simply based on mechanical aspects (e.g., Eigen value optimization), it is quite hard, in fact, to detect very uncommon design solutions or issues respect what already available as technical know-how such as, for instance, the superiority of reticular structures. Moreover, when further objectives are introduced in optimization, but always of the same type (such as, e.g., the minimization of an area keeping weight or stiffness unchanged), results do not change. Reticular structures seem the best solutions, especially when characterized by a certain geometrical complexity. However, as soon as a goal not strictly related to the 'structural engineering' is introduced, optimal solutions start to evolve along unpredicted directions.

For instance, a sandwich panel, made in composite by a single stratification that also considers the lighting holes, can be much simpler in terms of producibility respect to a composite grid. Then, when an additional parameter related to the producibility is introduced in the evaluation, moving the analysis from a 'topologic' to 'multi-object' approach, this perforated panel significantly increases its ranking respect to grid solutions. Similarly, lower density grids start to be more attractive.

However, without the possibility of objectively validating the weight of each objective, a deeper level of analysis risks to be inconsistent respect to the real applicability of results, which represents the essence and novelty of the present investigation. In fact, this study was intended to be a first attempt in the contest of solar vehicles to move from redesign action based, as tradition, on a trial and error approach toward a multi-step and -objective one method. To the knowledge of the authors, in fact, no other design studies are available dealing with the multi-objective optimization in an automotive context that involve fiber-reinforced sandwich structures of such large size and geometrical complexity.

4. Conclusions

The application of a multi-objective approach to the complete design process of an actual composite structure was shown. Specifically, a photovoltaic roof for solar vehicles was designed following a multi-objective approach in the way to balance divergence structural criteria, as static stiffness and dynamic response, with additional functional targets. Heat transmission and energy efficiency were also considered.

As first, several alternative shapes (circles, squares, triangles) were compared as pattern. Then, for selected shapes, the optimization was carried out respect to their main parameters (e.g., lengths, angles) searching the optimal points inside Ansys FEA software. A performance index was properly defined to represent the best compromise and a large number of configurations were compared. This index combined aspects as flexural stiffness, torsional stiffness, resonance frequency and heat transfer surface by the definition of weight parameters for each target. To simplify the study, a multistage approach was preferred. As first, a 500 × 500 mm section, equivalent to 4 × 4 solar cells, was adopted as base for an initial comparison between fundamental shapes (e.g., triangle, rectangle, ellipses). A total of 49 designs, each one characterized by a specific combination of shape and geometrical parameters (as widths, angles) were considered and compared, limiting the input for the next stage to few (4) optimal options. Hence, this base unit was used as design modulus to build a larger geometry (2500 × 800 mm) able to better predict and compare the structural behavior in a case study closer to reality. This second step of optimization, performed by 51 designs, permitted to recognize the quadridirectional grid as the best solution and to define proper combinations of geometrical parameters.

The FE model was validated comparing results from simulations and experiment respect to a modal analysis: a good accuracy with a 7% deviation in predicting the lowest natural frequencies was detected. The structure design, as here optimized thanks to a quadridirectional grid, was applied to the case of the real roof, characterized by larger dimensions (~5200 × 1.600 mm) and a double curvature geometry. A FE modal analysis of the roof, done in accordance with real loads and constrains, was carried out to determine the lower frequencies (higher than 12 Hz) and modes.

The composite structure was produced using autoclave technology; solar cells were also direct laminated on it. The solar panels, with an overall thickness lower than 5.2 mm, were installed on the vehicle and functionally tested on the road with valid results.

Author Contributions: Conceptualization, C.F. and G.M.; methodology, C.F. and G.M.; software, D.S. and A.P.; validation, A.P. and G.M.; formal analysis, A.P.; investigation, D.S. and A.P.; resources, G.M.; data curation, D.S. and C.F.; writing—original draft preparation, C.F. and D.S.; writing—review and editing, G.M. and C.F.; visualization, C.F.; supervision, G.M.; project administration, G.M.; funding acquisition, C.F. All authors have read and agreed to the published version of the manuscript.

Funding: This research has been carried out within the international collaboration project 'Two Seats for a Solar Car', an intervention funded by the Italian Ministry of Foreign Affairs and International Cooperation (MAECI), aimed at transforming a racing solar car into a solar road vehicle. The experimental activity was funded by the Italian Ministry of University and Research (MIUR) within the project PRIN2015 "Smart Composite Laminates".

Acknowledgments: The construction and installation of the composite roof were carried out by members of Onda Solare Sports Association at the MetalTig Srl workshops (in Castel San Pietro Terme, Italy). A special recognition for their commitment is dedicated to Mauro Sassatelli, Morena Falcone, Marco Berdoldi and Luigi Russi. The authors also thank to Marco Troncossi, Francesco Falcetelli and Alberto Martini for their support during the experimental sessions.

Conflicts of Interest: The authors declare no conflicts of interest. The funders had no role in the design of the study; in the collection, analyses, or interpretation of data; in the writing of the manuscript, or in the decision to publish the results.

Appendix A

This section contains details on the optimization procedure and parameters. The optimization was based on a three-step process where, as first, a selection of shapes was compared and optimized by 49 designs on a basic unit (500 × 500) (Tables 5 and 6 and Table A1), then, the comparison between geometries was improved by 51 designs at a level of mock-up (2500 × 800) (Tables 7 and 8 and Table A2), finally, the most promising solutions were evaluated in terms of multi-objective targets (Table 9). During the optimization, parameters were considered as fixed or modified inside the procedure.

Table A1. Preliminary (hole) shape optimization on basic unit.

Basic Unit	Parameter	Unit	Range	Type	
	Overall Dimension	mm	500 × 500	Fixed	
	Thickness	mm	5.2	Fixed	
	Holes distance	mm	125	Fixed	
Hole Shape	**Parameter**	**Unit**	**Range**	**Type**	**Simulations**
Rectangular	Width (*W*)	mm	50–95	Modified	15
	Height (*H*)	mm	50–95	Modified	
	Fillet radius (r)	mm	1–24	Modified	(Table 5)
Elliptical	1st axis (*A*)	mm	125	Fixed	9
	2nd axis (*B*)	mm	50–95	Modified	(Table 5)
Rectangular grid	1st Width (*a*)	mm	30–50	Modified	5
Quadridirectional Grid	1st Width (*a*)	mm	30–40	Modified	11
	2nd Width (*b*)	mm	14, 23.5, 25–30	Modified	
	Crosses		$1, 1\,^1/_2$	Modified	
	Spacing	Mm	250, 500	Modified	(Tables 5 and 6)
Quadridirectional Grid	1st Width (*a*)	mm	30	Fixed	9
	2nd Width (*b*)	mm	10.5, 20, 23.5, 28, 30	Modified	
	Number of crosses		$1, 1\,^1/_3, 1\,^1/_2, 2, 4$	Modified	(Tables 5 and 6)

Table A2. Grid optimization on the mock-up.

Mock-up	Parameter	Unit	Range	Type	
	Overall Dimension	mm	2500 × 800	Fixed	
	Thickness	mm	5.2	Fixed	
	Holes distance	mm	125	Fixed	
Hole Shape	**Parameter**	**Unit**	**Range**	**Type**	**Simulations**
Laminate Rectangle	Width (*W*)	mm	95	Fixed	1
	Height (*H*)	mm	93	Fixed	
	Fillet radius (r)	mm	23.5	Fixed	(Tables 7 and 8)
Quadridirectional grid	1st Width (*a*)	mm	30	Fixed	3
	2nd Width (*b*)	mm	14, 23.5, 30	Modified	
	Crosses Number	–	$1, 1\,^1/_2$	Modified	
	Spacing	mm	250, 500	Modified	(Tables 7 and 8)
Quadridirectional grid	1st Width (*a*)	mm	30–35	Modified	47
	2nd Width (*b*)	mm	14, 22–52.5	Modified	
	Spacing	mm	250, 375, 500, 625, 750, 875, 1000	Modified	(Table 8)

References

1. Tsirogiannis, E.C.; Vosniakos, G.C. Redesign and topology optimization of an industrial robot link for additive manufacturing. *Facta Univ. Ser. Mech. Eng.* **2019**, *17*, 415–424. [CrossRef]
2. Cross, N.; Cross, A.C. Winning by design: The methods of Gordon Murray, racing car designer. *Des. Stud.* **1996**, *17*, 91–107. [CrossRef]
3. Vadgama, T.N.; Patel, M.A.; Thakkar, D.D. Design of Formula One Racing Car. *Int. J. Eng. Res. Technol.* **2015**, *4*, 702–712.
4. Wloch, K.; Bentley, P.J. Optimising the performance of a formula one car using a genetic algorithm. In *International Conference on Parallel Problem Solving from Nature*; Springer: Berlin/Heidelberg, Germany, 2004; pp. 702–711.
5. Ustun, O.; Yilmaz, M.; Gokce, C.; Karakaya, U.; Tuncay, R.N. Energy management method for solar race car design and application. In Proceedings of the 2009 IEEE International Electric Machines and Drives Conference, Madison, WI, USA, 3–6 May 2009; pp. 804–811.

6. Minak, G.; Fragassa, C.; de Camargo, F.V. A Brief Review on Determinant Aspects in Energy Efficient Solar Car Design and Manufacturing. In *Sustainable Design and Manufacturing (SDM) 2017*; Campana, G., Ed.; Smart Innovation, Systems and Technologies; Springer: Cham, Switzerland, 2017; Volume 68, pp. 847–856.
7. De Kock, J.P.; van Rensburg, N.J.; Kruger, S.; Laubscher, R.F. Aerodynamic optimization in a lightweight solar vehicle design. In Proceedings of the ASME International Mechanical Engineering Congress and Exposition, Montreal, QC, Canada, 14–20 November 2017; pp. 1–8.
8. Odabasi, V.; Maglio, S.; Martini, A.; Sorrentino, S. Static stress analysis of suspension systems for a solar-powered car. *FME Trans.* **2019**, *47*, 70–75. [CrossRef]
9. Holmberg, K.; Andersson, P.; Erdemir, A. Global energy consumption due to friction in passenger cars. *Tribol. Int.* **2012**, *47*, 221–234. [CrossRef]
10. Betancur, E.; Fragassa, C.; Coy, J.; Hincapie, S.; Osorio, G. Aerodynamic effects of manufacturing tolerances on a solar car. In Proceedings of the International Conference on Sustainable Design and Manufacturing, Bologna, Italy, 26–28 April 2002; pp. 868–876.
11. Joost, W. Reducing vehicle weight and improving U.S. energy efficiency using integrated computational materials engineering. *J. Miner. Met. Mater. Soc.* **2012**, *64*, 1032–1038. [CrossRef]
12. Beardmore, P.; Johnson, C.F. The potential for composites in structural automotive applications. *Compos. Sci. Technol.* **1986**, *26*, 251–281. [CrossRef]
13. Nagavally, R.R. Composite materials-history, types, fabrication techniques, advantages, and applications. *Int. J. Mech. Prod. Eng.* **2017**, *5*, 82–87.
14. Elmarakbi, A. *Advanced Composite Materials for Automotive Applications: Structural Integrity and Crashworthiness*; John Wiley & Sons: Hoboken, NJ, USA, 2013.
15. Parkinson, A.R.; Balling, R.; Hedengren, J.D. *Optimization Methods for Engineering Design*; Brigham Young University: Provo, UT, USA, 2013.
16. Horst, R.; Pardalos, P.M. *Handbook of Global Optimization*; Springer Science Business Media: Berlin, Germany, 2013; Volume 2.
17. Yang, R.J.; Chahande, A.I. Automotive applications of topology optimization. *Struct. Optim.* **1995**, *9*, 245–249. [CrossRef]
18. Kim, C.H.; Mijar, A.R.; Arora, J.S. Development of simplified models for design and optimization of automotive structures for crashworthiness. *Struct. Multidiscip. Optim.* **2001**, *22*, 307–321. [CrossRef]
19. Cavazzuti, M.; Splendi, L.; D'Agostino, L.; Torricelli, E.; Costi, D.; Baldini, A. Structural optimization of automotive chassis: Theory, set up, design. In *Problemes Inverses, Controle et Optimisation de Formes*; Université Paris-Dauphine: Paris, France, April 2012; Volume 6.
20. Cavazzuti, M.; Baldini, A.; Bertocchi, E.; Costi, D.; Torricelli, E.; Moruzzi, P. High performance automotive chassis design: A topology optimization based approach. *Struct. Multidiscip. Optim.* **2011**, *44*, 45–56. [CrossRef]
21. Zhu, P.; Zhang, Y.; Chen, G.L. Metamodel-based lightweight design of an automotive front-body structure using robust optimization. *Proc. Inst. Mech. Eng. Part D J. Automob. Eng.* **2009**, *223*, 1133–1147. [CrossRef]
22. Diyaley, S.; Chakraborty, S. Optimization of multi-pass face milling parameters using metaheuristic algorithms. *Facta Univ. Ser. Mech. Eng.* **2019**, *17*, 365–383. [CrossRef]
23. Radiša, R.; Dučić, N.; Manasijević, S.; Marković, N.; Ćojbašić, Ž. Casting improvement based on metaheuristic optimization and numerical simulation. *Facta Univ. Ser. Mech. Eng.* **2017**, *15*, 397–411. [CrossRef]
24. Rao, S.S. *Engineering Optimization: Theory and Practice*; John Wiley & Sons: Hoboken, NJ, USA, 2009.
25. Yildiz, A.R. A new hybrid particle swarm optimization approach for structural design optimization in the automotive industry. *Proc. Inst. Mech. Eng. Part D J. Automob. Eng.* **2012**, *226*, 1340–1351. [CrossRef]
26. Nikbakt, S.; Kamarian, S.; Shakeri, M. A review on optimization of composite structures Part I: Laminated composites. *Compos. Struct.* **2018**, *195*, 158–185. [CrossRef]
27. Nikbakht, S.; Kamarian, S.; Shakeri, M. A review on optimization of composite structures Part II: Functionally graded materials. *Compos. Struct.* **2019**, *214*, 83–102. [CrossRef]
28. Muyl, F.; Dumas, L.; Herbert, V. Hybrid method for aerodynamic shape optimization in automotive industry. *Comput. Fluids* **2004**, *33*, 849–858. [CrossRef]
29. Ermolaeva, N.S.; Castro, M.B.; Kandachar, P.V. Materials selection for an automotive structure by integrating structural optimization with environmental impact assessment. *Mater. Des.* **2004**, *25*, 689–698. [CrossRef]

30. Minak, G.; Brugo, T.M.; Fragassa, C.; Pavlovic, A.; Zavatta, N.; De Camargo, F. Structural Design and Manufacturing of a Cruiser Class Solar Vehicle. *J. Vis. Exp.* **2019**, *143*, e58525. [CrossRef]
31. Pavlovic, A.; Fragassa, C. General considerations on regulations and safety requirements for quadricycles. *Int. J. Qual. Res.* **2015**, *9*, 657–674.
32. Pavlovic, A.; Fragassa, C.; Minak, G. Design and Analysis of a Carbon Fibre-Reinforced Plastic Roll Cage for a Solar-Powered Multi-Occupant Vehicle. *J. Sandw. Struct. Mater.*. under review.
33. Bogenfeld, R.; Kreikemeier, J.; Wille, T. Review and benchmark study on the analysis of low-velocity impact on composite laminates. *Eng. Fail. Anal.* **2018**, *86*, 72–99. [CrossRef]
34. Fragassa, C.; de Camargo, F.V.; Pavlovic, A.; Minak, G. Explicit Numerical Modeling Assessment of a Basalt Reinforced Composite for Low-Velocity Impact. *Compos. Part B Eng.* **2019**, *163*, 522–535. [CrossRef]
35. Rajbhandari, S.P.; Scott, M.L.; Thomson, R.S.; Hachenberg, D. An approach to modelling and predicting impact damage in composite structures. In Proceedings of the ICAS Congress, Toronto, ON, Canada, 8–13 September 2002; pp. 8–13.
36. Huybrechts, S.M.; Hahn, S.E.; Meink, T.E. Grid stiffened structures: A survey of fabrication, analysis and design methods. In Proceedings of the 12th International Conference on Composite Materials, (ICCM/12), Paris, France, 5–9 July 1999.
37. Dhir, S.K. Optimization in a Class of Hole Shapes in Plate Structures. *J. Appl. Mech.* **1981**, *48*, 905–908. [CrossRef]
38. Huybrechts, S.; Tsai, S.W. Analysis and behaviour of grid structures. *Compos. Sci. Technol.* **1996**, *3538*, 1001–1015. [CrossRef]
39. Januky, N.; Knight, N.F.; Ambur, D.R. Optimal design of grid-stiffened composite panels using global and local buckling analyses. In Proceedings of the 37th Structures, Structural Dynamics, and Materials Conference, Salt Lake City, UT, USA, 15–17 April 1996.
40. Mucchi, E. On the comparison between displacement modal testing and strain modal testing. *J. Mech. Eng. Sci.* **2016**, *230*, 3389–3396. [CrossRef]
41. Dos Santos, F.L.M.; Peeters, B.; Menchicchi, M.; Lau, J.; Gielen, L.; Desmet, W.; Góes, L.C.S. Strain-based dynamic measurements and modal testing. In *Topics in Modal Analysis II*; Allemang, R., Ed.; Springer: Cham, Switzerland, 2014; Volume 8, pp. 233–242.
42. Di Sante, R. Fibre Optic Sensors for Structural Health Monitoring of Aircraft Composite Structures: Recent Advances and Applications. *Sensors* **2015**, *15*, 18666–18713. [CrossRef]
43. Araki, K.; Ota, Y.; Yamaguchi, M. Measurement and Modeling of 3D Solar Irradiance for Vehicle-Integrated Photovoltaic. *Appl. Sci.* **2020**, *10*, 872. [CrossRef]

MDPI
St. Alban-Anlage 66
4052 Basel
Switzerland
Tel. +41 61 683 77 34
Fax +41 61 302 89 18
www.mdpi.com

Applied Sciences Editorial Office
E-mail: applsci@mdpi.com
www.mdpi.com/journal/applsci

www.ingramcontent.com/pod-product-compliance
Lightning Source LLC
LaVergne TN
LVHW070929170726
843515LV00005B/902

* 9 7 8 3 0 3 6 5 0 1 5 2 9 *